PC Mods for the Evil Genius

25 Custom Builds to Turbocharge Your Computer

Evil Genius Series

Bionics for the Evil Genius: 25 Build-it-Yourself Projects

Electronic Circuits for the Evil Genius: 57 Lessons with Projects

Electronic Gadgets for the Evil Genius: 28 Build-it-Yourself Projects

Electronic Games for the Evil Genius

Electronic Sensors for the Evil Genius: 54 Electrifying Projects

50 Awesome Auto Projects for the Evil Genius

50 Model Rocket Projects for the Evil Genius

Mechatronics for the Evil Genius: 25 Build-it-Yourself Projects

MORE Electronic Gadgets for the Evil Genius: 40 NEW Build-it-Yourself Projects

101 Spy Gadgets for the Evil Genius

123 PIC® Microcontroller Experiments for the Evil Genius

123 Robotics Experiments for the Evil Genius

PC Mods for the Evil Genius: 25 Custom Builds to Turbocharge Your Computer

Solar Energy Projects for the Evil Genius

25 Home Automation Projects for the Evil Genius

PC Mods for the Evil Genius

25 Custom Builds to Turbocharge Your Computer

JIM ASPINWALL

New York Chicago San Francisco Lisbon
London Madrid Mexico City Milan New Delhi
San Juan Seoul Singapore Sydney Toronto

The McGraw·Hill Companies

Library of Congress Cataloging-in-Publication Data

Aspinwall, Jim.
    PC Mods for the evil genius: 25 custom builds to turbocharge your computer / Jim
    Aspinwall. – 1st ed.
        p. cm.
    Includes bibliographical references and index.
    ISBN 0-07-147360-2 (alk. paper)
    1. Microcomputers-Upgrading.   2. Microcomputers-Maintenance and repair.   I. Title.
TK7887.AB2 2006
621.39'16-dc22   2006036983

1 2 3 4 5 6 7 8 9 0   QPD/QPD   0 1 3 2 1 0 9 8 7 6

ISBN-13: 978-0-07-147360-6
ISBN-10: 0-07-147360-2

The sponsoring editor for this book was Judy Bass, the editing supervisor was David E. Fogarty, and the production supervisor was Pamela A. Pelton. It was set in Times New Roman by Keyword Group Ltd. The art director for the cover was Anthony Landi.

Printed and bound by Quebecor/Dubuque.

This book was printed on acid-free paper.

McGraw-Hill books are available at special quantity discounts to use as premiums and sales promotions, or for use in corporate training programs. For more information, please write to the Director of Special Sales, McGraw-Hill Professional, Two Penn Plaza, New York, NY 10121-2298. Or contact your local bookstore.

Dedication

Of course, and she expects this, first and foremost, to my wife Kathy, an 'author's widow', who has been without a husband, the handyman, and my annoyances, for the past nine months while I shunned writer's block, the day job, California winter rains, and a record-breaking heat wave to crank out these modest words. No one but Kathy has any idea how long a honey-do list can be after so much time…

To my Mom and Dad, who I recently joked about still wondering what's wrong with me – I still have no clue!

Once again, to three very important influences on my PC and writing 'career' – Rory Burke for his initial PC mentoring, knowledge, and curiosity about things such as these; Mike Todd for his experience and insight into computing of all sorts and computer users; and Judy Bass, who has managed to tolerate every writers' excuse 'in the book', of which I'm sure she's heard and dealt with them all with humor, determination, and no less than positive encouragement to get it done. I can think of no one better to shepherd an author through the processes and get them to produce.

To Sally and the readers of the *Campbell Express* who I've left without my "Computing in Campbell" column for a few months so I could get this done.

To Robert Luhn, who still 'tosses me a bone' of contribution to pithy PC articles with unique technical challenges.

To 'MI', 'CK', 'SH', and the entire IT staff and thousands of our 'customers' at 'VMS' – you know who you are, and how crazy we get when things get 'fun' – it's not about us but the users we serve!

And, not in the least, to anyone and everyone who has dared read a word of what I've written in print or online and gotten even a little help from it. Technology is both challenging and stimulating and I do not envy any of us for trying to tackle it, but want to leave all with the encouragement that there are NO 'stupid questions' – keep asking!

Jim Aspinwall is the author of a number of PC-related books, was the "Windows Helpdesk" columnist for CNET, and has been a featured writer for PC World magazine. His articles were cover stories in both March and August, 2005. During the day he is responsible for the hardware and software configuration, updates, and general health for thousands of PCs world-wide. Jim is also an amateur radio operator, electronics technician, and OSHA-certified tower climber, maintaining a variety of radio sites in northern California. He is the author of Installing, Troubleshooting, and Repairing Wireless Networks (McGraw-Hill, 2003), as well as PC Hacks (O'Reilly, 2005), IRQ, DMA, and I/O (MIS Press, 1995), and several editions of Troubleshooting Your PC (MIS Press, 1994). He lives in Campbell, California.

Contents

Dedication   v

About the Author   vi

Contents   vii

Preface   ix

Acknowledgments   xi

Introduction   xiii

1 The Green PC   1

2 PC Control Panel   5

3 The Ultra-Mod PC   9

4 Motherboard Upgrade   17

5 Build Your Own Power Protection   23

6 Secure Your WiFi   31

7 Dynamic Picture Frame   39

8 Mobile Entertainment   51

9 Mobile Navigation   59

10 Mobile Internet   65

11 Share Your Trip With APRS   69

12 Let Others Track Your Trip By Google   83

13 PC Weather Station   89

14 I Want My PC TV   97

15 Sling Video All Around   101

16 Voice-Over-IP With Skype   109

17 Security Webcam   115

18 Contributing to Science Explorations   121

19 Keep In Sync   127

20 Securing Your Computer   135

Glossary   143

Index   171

Preface

What?

Somewhere, right now, hundreds of people have the covers off their PCs, exploring the insides, upgrading a video adapter, disk drive, or adding memory or a DVD drive. Those are the people most consider to be the experts, hackers, geeks, or nerds who 'know everything' about computers. I'll let you in on a little secret one of my PC experts taught me – 'no one can know everything'. To go with that, 'an expert is someone who knows just a little more than you do'. Soon you will realize that you can become an expert at something your favorite expert has not yet accomplished.

PC Mods for the Evil Genius is about many of those special things you'd like to be able to do with a PC but may not have an expert to guide you. Every project in this book is doable by those of us who have not done anything like this before – from customizing the appearance of a PC to suit a room or a theme much like the designers and craftspeople do to houses and rooms on *HGTV*, to adding the latest in high-tech security to keep friends, family or coworkers from messing with your PC.

Every one of the mods presented to you in this book was inspired by my own desire to know, do, play, have a PC do something I know it should be able to do, obtain some new functionality, or simply have fun. I knew these mods were possible, some of them lingered on the back-burner of my perpetual, endless task list for months, years. Now was as good a time as any to get them done. If they inspire you – so much the better!

Do I have a favorite mod? No, I have several. Creating a GPS-driven mobile navigation and tracking system is perhaps the coolest challenge/accomplishment of them all. Dressing up a PC for my granddaughter allowed me to pass on a comparable gift from my grandfather when I was younger. Turning a well-used laptop into a dynamic work of electronic art actually impressed my wife and friends while it got me away from the keyboard and into the shop for a few minutes. The more you think about doing, the more you will try to do, the more you will accomplish.

Why?

PCs and their associated gadgets exist for two reasons – because people want to be able to do something of interest to them with their PCs, and because PCs can do just about anything. When you get your PC to do something that is special to you, you've become an expert on that 'something'. Getting your PC to do a certain 'something' is rarely well-documented in the instruction manual, magazine review or shared by friends or other experts. My goal is to present a lot of common, cool things a PC can do, that you might enjoy and be inspired to explore further.

This book is not about hacking, cracking, tweaking, super-charging, over-clocking, or bending the rules of hardware, software, or the Internet. I've written about that. When presented with this project my mind immediately went to a classroom full of 13- and 14-year-old computer enthusiasts at Branciforte Junior High in Santa Cruz, eager to jump into writing games programs, wondering if they could hack the school's computer to alter their grades, or simply be able to spend time surfing the Internet. Then I looked around my home 'office'/lab/workshop and fondly remembered high school shop class and all I learned then and enjoy today that makes me one of those so-called experts with PCs. Anyone who has seen my 'office' and shop knows immediately that I am a bit nuts, and while neither evil nor genius (perhaps 'Nutty Professor'), some pretty neat things come out of my head and eventually off my workbench.

In my career working with both technical and non-technical people, and more often than not, I was *not* the expert and needed to learn. I know the sense of enablement, entitlement, ability, and reward when something new is accomplished. Teaching my high school 'girlfriend', a true technical novice, how to fix radio pagers for doctors 30 years ago gave us both a sense of accomplishment – she learned a new skill, I realized I could help someone else learn and do more.

As I encountered the world of computers and then PC hardware and software, I met and appreciated a new level of experts. Suddenly everyone wanted 'in' on the new world of information technology, the Web, programming, databases, servers, and routers. I saw that more and more people came to this technical field with no technical background at all – they got a PC for Christmas, got a dial-up account with AOL or Earthlink, went to school, took some programming classes, and became 'experts' seemingly without knowing how or why their PC or the Internet really works.

These are the people who make the hardware and software products we have to put up with, make tech support calls about, and may eventually return to the store because we cannot figure out how to make them work. Failure of 'simple' technology is unacceptable in my book – we will get things to work!

I want to make sure that the people who touch PCs have a much better understanding and appreciation for what a PC is, why PCs are the way they are, what PCs can and cannot do, how they do it, and how to do it better, easier, faster, less expensively. Being a geek, 'engineer', 'expert', product designer/inventor of sorts myself, it is easy for me to see-through various products, envision how and why they work, and when the moment is upon me, replicate them in my own way, and save a few dollars at the same time.

Some of these mods may not appeal to you for home use, but may be invaluable to your local school's science or computer class. We're all curious about and affected by the weather, and wireless digital communication lives in just about everyone's pocket, so there are many opportunities to inspire, entice, create, and benefit.

I am anxious to learn of your success stories, and especially of any enhancements you might make to any one of these mods. Nothing would please me more than to see one of my readers take first place, get a blue ribbon, or add a special skill to a resumé that sets individuals apart from others through one of these mods.

How?

You can accomplish these mods with a quick trip to the hardware store, a few minutes downloading from the Internet, grabbing a gadget or two from the local computer shop, but always in the comfort of your home or someone else's handy workspace. Most of these mods take no more than an hour or so from start to finish.

Mods that involve serious tool work to paint, drill, etc. will take a bit longer, and adult supervision if tackled by younger evil geniuses. If you lack a tool or skill to accomplish some of the more involved mods, likely there is someone in a computer, model, robot, or amateur radio club who would be more than willing to help.

If you encounter a question, difficulty, glitch, error, conflict, or some other challenge that gets in the way of accomplishing any of these mods, my goal is accomplishment and I will support you within reason – directly or pointing you in the right direction. I have a motto at work: 'the machine never wins.' Software and hardware vendors rarely deny an opportunity to have their product work, and I am no different – whether it's my 'invention' or I need to call a vendor for help.

Let's have some fun and do some crazy stuff with our PCs!

Jim Aspinwall

Acknowledgments

There are always many passive and active contributors we draw upon to create such work. In this book I've called upon the lessons of just about every teacher, product, discipline, mentor, and idea one can imagine when it comes to PCs, electronics, mechanics, tools, software, and PC hardware, and providing me with a tolerable foundation in literacy and expression to share ideas and their execution. I would like to specifically acknowledge:

- My Uncle Mike, a firefighter and good person, and Grandpa Bill who was fair to us all, who I didn't know at the time they were with is, would be inspirations to tackle just about anything, and do the simplest yet impressive things with a saw, some good pieces of wood, a quart of paint, and love.

- Bernie H and Mike H – my electronics and metals shop instructors in high school – there is everything good in learning the basics of electrons and tools – sometimes we actually do something with what we've learned!

- The vast array of people in amateur radio, especially those who made 'packet radio' and 'APRS' happen for us – some truly innovative work that does and will attract many 'digital age' inhabitants to our amazing hobby.

- 'TM' of Microsoft for providing "Microsoft Streets and Trips with GPS" – of all the GPS and mapping programs I've explored this past year 'S&T' has made the journey through this book and summer vacation enlightening, and I dare say, it is also very 'easy to use'.

- The folks at Slingbox for a truly wonderful product – have I got some ideas for you folks!

- The folks at DigitalPersona for U.are.U – ditto – we should talk about biometrics and Windows security! Thanks for the product!

- 'CyberGuys' – their catalog is like "Popular Science" or the "Heathkit" catalog for the 21st century – your products inspired the 'frog' PC, among other projects.

Learning, and sharing what we know, what we find out, and what we can do with it all, is a tremendous opportunity for those of us who translate technology into words and pictures, and best of all, success, productivity and fun for those who partake of our services. If it's not fun for both us and the people who use this stuff, we're doing the wrong thing!

My working title for this book has been "PC science projects." That became my guide in drawing out interesting things to do with PCs in a variety of environments. I fondly remember a room full of 13 to 14 year-olds at a junior high school in Santa Cruz who had an insatiable appetite for all things PC (no, not 'politically correct') when the so-called experts came to present to the computer club. I find that many adults are no less fascinated by the cool things that can be done with a computer.

I've tried to keep the projects safe and doable by most anyone, even if some of the more serious parts required adult supervision – though I suspect some kids are more aware and careful than some adults.

I suspect that many of this book's readers have been raised in or have been through the era of technology explosion – perhaps not as far back as when color TV or even TouchTone™ phones were new, but certainly people who vividly remember the IBM-PC/XT, modems, CompuServe, e-mail, Windows 3.1, and Netscape 1.x. The nuevo-techno generation, or their offspring known today as 'Generation Y' – grandchildren-age to some of us – may never see a 'shop class' in their secondary education, and think learning how to use a hammer or drill happens only on Saturdays at Home Depot.

Intel, IBM, COMPAQ/HP, Gateway, and Dell, as well as AOL, Netscape, and Microsoft, have created this wild world of PCs and opportunities to do some tremendous things with them. WWW, HTTP, RSL, Yahoo!, Google, 'Moto', MySpace, 'texting', 'blogging', and downloading 'Top 40' ring-tones are the world of technology most people know today – services and features that require hard-core technology and smarts to navigate around. All of these technologies end up at our desks, in our laps, or in our pockets. Provided we are curious and adventurous enough, we can spin them into some

fantastic awareness and experiences we can relate to and use to enhance some aspect of daily life.

It is hard for most of us to imagine a world without e-mail, Yahoo!, Google, or ubiquitous high-speed Internet access; remote access to any PC around the globe or playing with a web-cam above Fisherman's Wharf in San Francisco. It is especially hard for me to imagine anyone with an interest in 'technology' who is not willing or able to open up the case of a PC to add a disk drive, change out memory chips, or install a new I/O card – but then, that is the challenge presented to us 'experts' – to articulate, encourage, enable, and support anyone and everyone who owns a piece of high-tech gear to dive in and make the most of it.

I perpetually want to share the enlightening, inspiring moments of discovery and accomplishment that empowers any of us to do more with the tools at hand. Such is the driving force of this book – to illustrate, inspire, encourage, enable, and empower anyone and everyone to take a calculated chance with their PC and individual skills to go beyond e-mail and web-browsing.

What you will see in the next 20 chapters are doable, exciting, beneficial projects for home, school, computer club, and daily applications. Writing this book has been a good opportunity to complete many of the projects I've had on the to-do list for many months.

Among the many, many things that can be done with a PC, my favorite projects or 'mods' so far, for a variety of different reasons have been:

- The 'green PC' for kids – inspired by CyberGuys for my granddaughter.
- Mobile navigation – because knowing where we are is part of getting there.

- Trip-tracking with Google Earth – a vagabond blog of sorts.
- Fingerprint security – it's a high-tech push button.

As commercial and enterprise computing experts talk about 'convergence' of voice and data services, many in the computing and radio hobbies continue to explore voice-over-IP-over-radio in various forms to unite the globe in the only truly open-source, cross-technology realm that exists in amateur radio.

I've been known for my knack with troubleshooting and fixing technical problems. I've devised and fostered new software to help people with PCs. I've written five books about PCs, but none has been as challenging (in terms of my obligation to you, the audience), anticipated, and hopeful as the current book in illustrating some pretty amazing PC tricks.

I am and have been since childhood, an amateur radio operator. Through my career I have taken that hobby beyond anything resembling 'amateur' status, but embrace the opportunity to experiment, tinker, learn, build, and do useful things with all sorts of technologies. It is my desire, my hope, my goal to have each of my readers take something inspiring, beneficial, valuable, and perhaps even life-altering away with my work – find or expand a career interest, take on a new major in college, acquire a new merit badge in scouting, get your amateur radio license, or simply have something cool to show off on your PC.

The phrase 'PC mods' provides a lot of opportunities – this could be a book about altering or making some truly impressive PC packages – from monster themes to bunny rabbits. This could be a book about hacking into PC, video card, network adapter, or disk drive firmware to add or alter features. This could be a book about upgrading your PC to the next level of performance or preparing for the next operating system.

I chose to make the 'mods' in this work include approachable and interesting options from the outside moving inward. From case to cameras, power cords to password replacements, just about anything you can do with a PC is here.

I will admit that this book lacks two very compelling 'mods' due to time, space, and complexity – a completely PC-controlled robotic project, and a very powerful PC-driven machine tool. I have yet to find a robot that has a PC on-board as its 'brain'. Indeed many special-purpose embedded computers for robots are 'fed' their programs from a PC but none of the robot projects I've found has a working PC running Linux, DOS, or Windows on-board. The machine tool – essentially an X-Y plotter that moves a cutting tool across a wood, circuit board, or metal object – is very, very interesting, but the focus is more on controlling the movement of the tool rather than on gaining an exceptional benefit from the PC alone.

Another missing project, which I am leaving up to Microsoft, Gateway, Dell, Compaq/HP, or some other PC vendor to work out, is a full-featured media center PC – one that takes all of our VCRs, DVD players, cable, and over-the-air antenna inputs, and allows us to 'route' these inputs to other TVs or recorders throughout our homes. The complexity with this is not so much the PC but the lack of 'routing' equipment to control which signal goes where – and adapt antenna signals to S-video or component, etc.

I hope you will forgive what I am lacking in this work and enjoy what is offered for your enjoyment and benefit. Each and every one of these 'mods' is alive and running in my home, car or, office and provides a lot of enjoyment and benefit. Let me know your favorites and your ideas ... meanwhile – happy bits!

Jim, aka, pcmods.evilgenius@gmail.com

Chapter 1

The Green PC

My granddaughters get most of the credit for inspiring this project, but if you consult any child you will probably receive some terrific ideas towards creating a similar project to suit a child's favorite color, animal, cartoon character, super-hero, or food.

We're a fairly tech-savvy family so creating a customized PC for a two-year-old is not that far out of reach around here, though you may find this mostly applies to kids of five and over – not the building, but enjoying the results.

Grand-daughter Kimberly likes frogs and dinosaurs, while Renae loves to track down the lizards in our backyard, so I was drawn to create this PC project when I saw a couple of key components offered for sale at www.cyberguys.com – specifically a frog-green mouse with roving eyes and a matching green keyboard. My wife may have preferred that I create a pink 'Barbie PC' but I couldn't find a pink keyboard and figured the tiny frilly decorations and glitter typical of 'Barbie' decorations and accessories would mess up the works. Having a theme, base color, or key piece to start with drives the rest of the project, just like redecorating or remodeling a home. With the main theme established you can decide if painting or some other treatment is right to bring all the pieces together. In this case all that was needed was a new paint job for the computer case, its components, and an LCD monitor to match.

Preparation for this project begins with scrounging or buying a PC case that can be disassembled down to the buttons and springs and faceplates. Broken down, the components of the case can be repainted in different colors to provide some really elegant touches.

Figure 1-1   *The green frog mouse from Cyberguys.com. The eyes are the left and right buttons.*

Make sure you get a case that will fit the motherboard you buy or choose to use. I started with a system that contained an 'old' Intel Celeron board in ATX form-factor so I could upgrade the insides later on. Also look for peripherals – CD-ROM and diskette drive – that allow you to remove their faceplates so they can be repainted.

I don't recommend you attempt disassembling and repainting a mouse or keyboard – I've done that before to create a theme for trade shows – it's not fun. If you're 'going green' as I did – obtain the green keyboard and mouse from www.Cyberguys.com – they also offer a variety of accessories in other colors. Unfortunately this book is in black-and-white so you'll miss most of the transformation from beige to green but the steps are all here.

CAUTION! This project involves the use of hand tools, common shop chemicals, and aerosol paint products. Adult supervision, adequate ventilation, and proper eye and respiratory protection are recommended.

Tool Kit

- Spray paint – your choice of type and colors. For the painting you may prefer to use a nontoxic or latex-based spray product instead of the typical Krylon or Rust-oleum found in most hardware and building supply centers.

- Flat-blade and cross-point screwdrivers.

- Masking tape.

- Sharp knife (X-acto™ or razor-blade type).

- Newspaper or other material to be a drop-cloth/ back-drop for spray painting work.

- Shop rags – old small towel, paper towels or similar.

- Soap, water, glass cleaning solution.

- Laquer thinner.

Parts List

- PC case that can be dis-assembled for repainting.

- CD-ROM drive that front panel can be removed from for repainting.

- Diskette drive that front panel can be removed from for repainting.

- Optional: CRT or LCD display to be repainted.

Steps

1. Disconnect all cables and power cords from PC and display components.

2. Remove the top cover from PC case. Some cases may have up to three covers – two sides and a top.

3. Remove front panel from PC case – typically by unsnapping the plastic latches or in some cases removing screws that hold the panel in place.

4. Detach indicator LEDs and buttons from front panel.

5. Remove faceplates from CD-ROM and diskette drives.

6. Clean the surfaces to be painted first with soap and water or household glass-cleaner and let dry thoroughly. Wipe down the metal surfaces to be painted with a shop rag dabbed in a bit of lacquer thinner to prepare the original painted surface to take a new coat of paint.

7. Apply masking tape to any areas of case, display, front panel, or drive faceplates you do not want exposed to paint. Pay particular attention to electrical connections (power, data, video, USB, etc.) and knobs, buttons, and switches that could be overpainted, and stick or be blocked from excess paint.

Figure 1-2   *Front cover of PC detached from chassis with LEDs and switches removed from their brackets.*

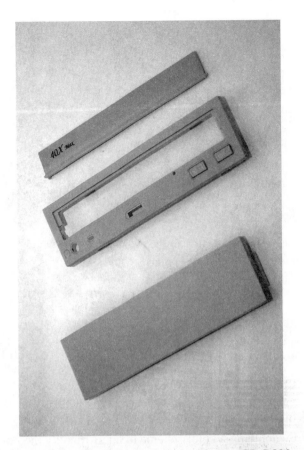

Figure 1-3   *Removing the faceplate from a CD-ROM drive.*

8. Spray paint case components and monitor to suit the color scheme you desire.

9. Follow good techniques for painting – 2–3 lighter coats produce better results that trying to cover the work area with one coat. Let each coat dry 2–4 hours before applying next coat and before beginning to reassemble the pieces.

10. When the painted pieces are dry, carefully remove masking tape from all taped pieces.

11. Reassemble front covers of CD-ROM and diskette drives.

12. Reattach the LEDs and button pieces on the front panel.

13. Attach front panel to PC case.

14. Replace the cover over PC case.

Figure 1-4   *Covering the viewable area of an LCD display with masking tape before painting.*

15. Connect custom mouse and keyboard.

16. Connect the video display to the PC.

17. Reconnect power to PC and video display.

18. Start the system, make sure everything works as before and enjoy.

Figure 1-5 *Removing the masking tape from the electrical connections of the LCD display.*

Figure 1-6 *The completed green PC.*

Summary

In this project we've completely redecorated a PC by applying a color scheme to a kid-friendly theme sure to last for a couple of years. This type of project, though it does nothing to enhance the performance or capability of a PC, lends itself well to everything from a popular super-hero to NASCAR to 'Barbie' or similar popular themes.

As you get more comfortable with PC components your child's imagination and your skill level will expand to create myriad possibilities and customized PCs.

Chapter 2

PC Control Panel

PCs are often pretty mysterious and downright inconvenient to work with at times. It's taken years for PC makers to figure out that there are a lot of widgets we need to plug-in and unplug on a regular basis – especially now that we have FLASH memory 'sticks' in cameras, cell phones, and as replacements for diskettes and CDs. We also wonder sometimes if the odd behavior of our PC is because it's too hot inside, and at the same time are annoyed by how noisy these contraptions can be.

To help us out of these dilemmas I found an economical accessory to bring a lot of frequently used ports and illustrating controls to our fingertips – a multi-function PC control panel, shown below.

A control panel begins to make sense as the perfect add-in accessory when you look around your desk and see a card reader for your cell phone and camera memory devices, a USB hub, and probably an extension cord from your PC speaker jack for your earphones – all with cables that get tangled in your keyboard and mouse wires as they wind their way to the back of your PC.

This project helps end the techno-clutter on your desktop as well as telling you the temperature inside your PC's case and whatever else you'd care to monitor – typically keeping an eye on how hot the CPU is. Your PC must have an empty 5-1/4-inch disk drive bay in which to place the control panel. The complexity level of this project is comparable to installing new disk drives, but without the formatting and software hassles, and it doesn't change the way anything works – it simply adds functionality.

Tool Kit

- Various sizes of flat and cross-head screwdrivers to suit the screws in your PC.

- Wire cutters (to clip the ends off of wire ties).

Parts List

- Sunbeamtech 20-in-1 Superior Panel (www.cyberguys.com).
- Assorted wire ties.
- An extra set of disk drive rails and appropriate screws to mount and secure the control panel inside your PC. (Many PCs come with extra mounting rails stored inside the case.)

TIP: Give yourself plenty of room to work on this project as you will need to open the PC chassis and access the front, rear, and insides.

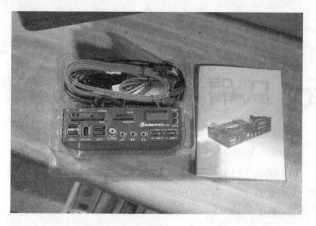

Figure 2-1 *The Sunbeamtech control panel adds several useful I/O ports, temperature display and fan speed control to most PCs.*

Figure 2-2 *Removing a filler panel from an empty drive bay.*

Project Steps

1. Power down and disconnect all cabling from your PC.

2. Place the PC on a surface with plenty of room to work.

3. Remove or open the outside cover from the PC chassis to gain access to the inside.

4. Remove the blank filler panel (Figure 2-2) from the front of an empty 5.25 inch wide disk drive bay.

5. Determine and assemble the proper rails or other mounting hardware to secure the panel inside the unused drive bay. *Do not mount the control panel in the bay at this time.*

6. Layout the cables (Figure 2-3) and check the instructions for the control panel to familiarize yourself with and plan the proper connections.

TIP: USB, fan and audio connections vary from PC to PC. In some cases you can access a USB port only at the rear panel rather than be fortunate enough to have an internal plug for extra USB devices. For fan wiring you may have to cut and splice various wires – not a difficult task, but if

your electrical skills end at working only with wires that have pre-attached plugs on the end, you may want to bypass any tricky rewiring steps or enlist the help of a friend.

7. Consult the manual for your PC's main board (a.k.a. motherboard or systemboard) to determine the availability and location of USB and fan connections you want to use for the control panel.

8. Connect the appropriate cables to the control panel and lay them out neatly

Figure 2-3 *The control panel and assorted interconnection cables to be sorted out and plugged into the panel and your PC.*

Figure 2-4 *Selected cables attached to the control panel ready for installation in the PC.*

Figure 2-5 *Installing the control panel into the drive bay from the inside of the PC.*

Figure 2-6 *Align the front of the control panel with the front of your PC before securing.*

Figure 2-7 *Arrange new wires and tie them to existing ones.*

(Figure 2-4) so you can arrange and bundle them to the proper locations inside your PC. Leave the bundling and putting wire ties around the cables for Steps 10 and beyond.

9. Insert (Figure 2-5) and align the face (Figure 2-6) of the control panel in the empty drive bay. Once aligned, secure the mounting with screws at the side of the drive bay.

10. Route the cabling within the PC chassis and connect to the appropriate I/O ports you'll be using (USB, SATA, IEEE-1394, audio, etc.) With the wires routed and connected you may begin to bundle and wire-tie them. Follow the routing of other wires to make a nice package. Allow sufficient slack to clear items inside and allow you to close the case without pinching or disconnecting the wires.

11. Insert one of the temperature sensors, 'T1', (Figure 2-8) between two fins of the CPU heatsink for indirect CPU temperature measurements.

12. Attach the 'T2' temperature sensor (Figure 2-9) to a convenient location near the top of the PC for chassis temperature measurements.

TIP: Mounting the chassis temperature sensor near the ventilation intake holes of the power supply will give a pretty accurate reading of the highest air temperature inside the case. The surface of disk drives, video, and memory chips may be much warmer than the aggregate inside air temperature.

Measuring the overall air temperature and comparing it with the room temperature is a good indication of how well your PC ventilation works. For instance, if the room temperature is 72 degrees

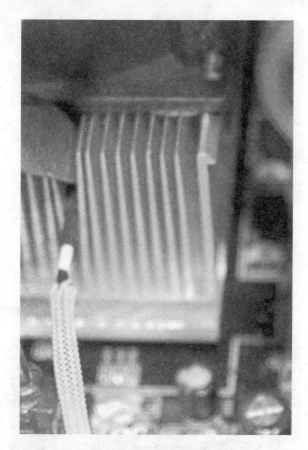

Figure 2-8 *The temperature sensor for the CPU fits nicely between fins of the heatsink.*

Figure 2-9 *The chassis temperature sensor works well near the power supply intake vent.*

Figure 2-10 *The completed control panel showing CPU temperature with a USB FLASH memory drive in use.*

and the temperature inside the PC is 90 degrees, your PC is ventilating well. If the temperature inside the PC is 100 degrees or hotter your PC needs additional ventilation.

Summary

The steps and techniques in this project are typical of many PC component upgrades, modifications, and accessory installations. Adding functionality without affecting how things work already is a great confidence-builder and prepares you for more radical steps ahead.

Chapter 3

The Ultra-Mod PC

As kids grow up, so do their tastes – from super-heros and cartoon characters to the gothic, arcane or super-techno. Even as adults many of us never lose our fascination for things that shine and blink and tell us how fast or strong or elegant something is.

For this project we leave one PC fantasy-land for another by way of 'shop-class'. We outgrow frogs and turtles and frilly things and shift into raw power – or at least the appearance of power. The techniques from Chapters 1 and 2 come into play, adding work with a few common power tools and some imagination.

The goal of this PC project is to create another new style of techno-PC using basic techniques and components for cooling, lighting, case decoration, and just plain obscure customization. You can let your imagination run wild – from a skillfully crafted color scheme to rocket-ships, or retro art. If you are into 'The Matrix' you could just strip a PC and display and toss components together into some form of world-after-nuclear-blast motif – but that brings up safety concerns not worth the hassle.

For those of you too young to remember, much less know about 'shop class', it was one or many elective courses providing skills education in wood-working, metal-working, auto-repair or electricity/electronics. 'Shop class' is where many 'old' folks learned handy skills such as those used throughout this book, and from there grew up into adult nerds or even hackers of sorts.

TIP: The activities we'll use here involve drilling, cutting, and painting a metal PC case, and making some internal wiring connections for fans and lights. The drilling and cutting should be performed or supervised by an adult, and anyone near the work in progress should have eye protection and exercise safety precautions.

Tool Kit

- Common hand tools – screwdrivers, pliers, wire cutters.
- Common power tools – drill motor with bits, jig saw with metal cutting blade.
- Ruler and pencil or marker.
- Spray paint.
- Shop rags.
- Drop cloth or newspapers.

- PC case suitable for modification – preferably one with separate side panels rather than one large inverted-U style cover. You may buy a 'mod' case ready to use or create your own from the classic off-the-shelf beige variety.

- CD-ROM/DVD and diskette drives with removable faceplates for customization.

- New power supply, motherboard, CPU, and memory if building this system from scratch.

- Control panel from Chapter 2.

- Multicolored cabling for disk drives, fans, power leads.

- Multicolored expandable cable covering sleeves.

- Colored wire-ties to suit your theme.

- Custom thumbscrews for case decoration/assembly.

- Fan with embedded LEDs to suit color scheme

- Cold-cathode lamp(s) for internal lighting effects.

As you have probably noticed there are a lot of different tools, pieces and parts to work with in this project. You'll want to allow 2–3 evenings or a full weekend to complete this project if you're doing significant modifications, or 1–2 evenings or a Saturday using a pre-fab 'mod' case. I started with a new classic beige box and worked from there.

Whether you assemble, test and then disassemble all of the pieces to ensure everything works before you start, or do the case modification work first then assemble and test everything for the first time is really up to you. My preference is to ensure all the components, at least the basic PC, works before committing a lot of time to modifications and new features.

For this project the motherboard, power supply and disk drives have been tested together and are known to be working – they've simply been removed from service in a different case and are being moved to this new case after modification. You could as well 'bench test' the critical components together outside of the case just to be sure.

Project Steps

1. Disassemble case. This means totally stripping all switches, LEDs, brackets, buttons and 'decorations' from the PC chassis – Figure 3-1. If you are modifying a working PC, you must also remove the disk drives, system board and power supply.

2. Remove the face plates from disk drives.

3. Carefully paint the chassis and drive parts, saving any modified/cut pieces, as in Step 4, for last. I chose to use a glossy enamel paint

Figure 3-1   *PC front panel parts separated for painting.*

Figure 3-2   *CD-ROM and diskette drive parts separated for painting.*

Figure 3-3   *Stripped down PC case being painted.*

Figure 3-4   *Drawing the location of the fan and marking the holes for mounting.*

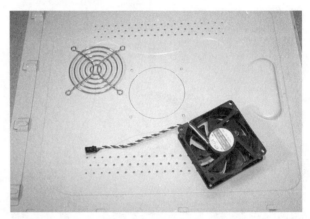

Figure 3-5   *Drawing the outline of the grille as a guide for cutting its circular hole with the jig saw.*

on the metal pieces of the chassis and disk drive faceplates, and a glossy lacquer for the PC case front and buttons. It takes 3–4 light coats of paint to properly cover all of the pieces – allow 2–4 hours drying time between coats and before re-assembling the parts.

4. Layout the location for any additional fans or accent holes you wish to add to the sides of the case. Here I chose to add a multi-colored fan to the main 'service' cover for the PC and matching vent holes and grille on the opposite panel.

TIP: When selecting the location for adding internal parts, always check to be sure the new part will fit properly and not collide with the power supply, disk drive bays or other components of the assembled PC.

Measure and mark the location for the fan, Figure 3-4, and the outline of the external grille, Figure 3-5, on both side panels. When you have the circle laid out, measure in 3/16 of an inch and mark a spot for a 3/8-inch pilot hole for the jig saw blade. Verify the location will not block or touch other parts and allows you to install and remove the cover.

5. Drill the four fan and grille mounting holes – typically 5/32-inch to use #6-32 screws.

6. Using a jig saw with metal-cutting blade, Figure 3-6, cut out the circular hole for the fan grille.

7. Clean metal burrs from the grille cutout with a fine file or emery cloth and check the fit of the grille, Figure 3-7, on both side panels.

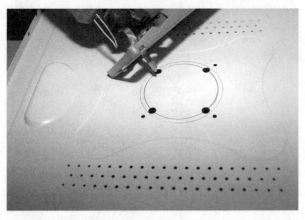

Figure 3-6   *Side panel ready for cutting circular hole for fan and grille. Note the pilot holes inside the area to be cutout.*

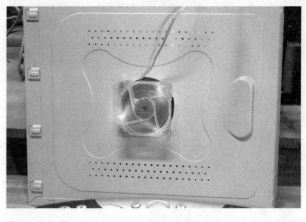

Figure 3-9   *Testing new fan.*

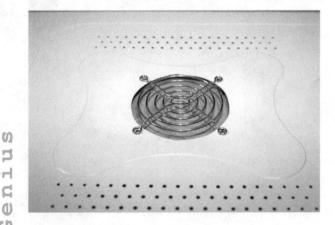

Figure 3-7   *Finished grille hole cutout and cleaned.*

8. Paint the side panels then assemble the grille and fan mountings – Figures 3-8 and 3-9.

9. Re-assemble the case and disk drive parts. Make sure all of the buttons operate

Figure 3-10   *Re-painted case with motherboard re-mounted.*

smoothly. Lightly sand the sides, edges and holes for any button that sticks.

10. Install the motherboard on its mounting plate. Be sure every mounting hole has a stand-off and screw to secure it in place. Do not place mounting standoffs where there is

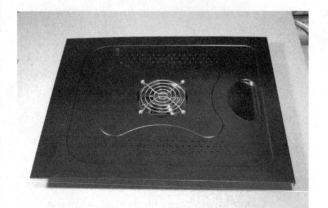

Figure 3-8   *Painted side panel with fan and grille mounted.*

Figure 3-11   *Case with disk drives and power supply re-installed.*

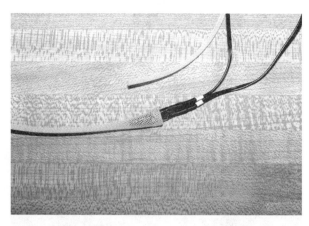

Figure 3-13   *Inserting a thermal sensor into expandable sleeve.*

not a matching hole in the motherboard. Install the motherboard into the chassis, Figure 3-10.

11. Install the power supply. For this project, rather than obtain a custom power supply I simply painted the exposed back side to match the case (masking off the switch, electrical plugs and fan hole before painting).

12. Install the remaining disk drives, Figure 3-11. Note the control panel from Chapter 2. This completes the main chassis assembly so you can begin wiring things together.

13. Prepare the wiring. For this step I selected a bright set of glow-in-UV-light cables and cable sleeves, Figure 3-12, to be shown off by a bluish cold-cathode lamp inside the

case. PC mod 'purists' will tell you that using round cables instead of flat ribbon cables for common disk drive connections helps improve air-flow throughout the inside of the PC case – I'm not sure it matters much but they provide a lot cleaner appearance.

Applying the expandable sleeves to the cables is a bit tricky, Figure 3-13 – but since the sleeve is expandable it's just a matter of protecting the ends of the cables or connectors from snagging on the sleeve and then pushing them through.

Figure 3-14 shows all of the cables dressed-up, ready for new connections to be made then bundled into a neat package. Notice the disk drive is connected – so I could test the system before all the new connections are made.

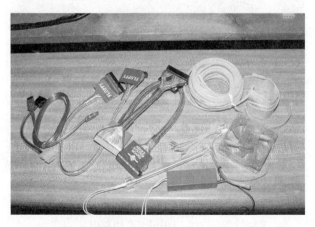

Figure 3-12   *An assortment of new cables and cable sleeve to be installed.*

Figure 3-14   *Cabling dressed waiting to be connected and bundled.*

Figure 3-15   *Inside the new case with new wire dressings lit-up under test.*

14. Neatly dress and bundle the cables, Figure 3-15, so the system can be completely tested. This is a considerably nicer layout than having a blob of crinkled ribbon cables congesting the layout.

Figure 3-17   *Rebuilt modified PC setup and running.*

Figure 3-16   *Rear-view of the modified PC with internal to external cable connections.*

Figure 3-18   *The finished project, complete with black Logitech cordless keyboard and mouse, will go together very nicely with a black LCD display panel.*

15. Connect appropriate cables to the rear of the chassis, Figure 3-16. Notice that the audio and USB connections for the front control panel are taken from the original connections at the back of the motherboard.

16. Complete the assembly with the covers, Figure 3-17, to see if you've got the desired 'mod' effect, and admire the finished project, Figure 3-18.

Summary

This project obviously involves a lot of different tasks, from painting to metal-work to wiring. Similar projects have been done building PCs into small refrigerators, old wooden radio cabinets, microwave oven chassis, all the way to the extreme of creating fiberglass and plastic creatures to contain the working parts of a PC – a few examples are available at sites like www.creativemods.com. If you do not have the facility for all of the painting and cutting you can buy pre-modded cases from vendors such as www.thermaltake.com, www.xoxide.com, www.frozencpu.com, and www.crazypc.com.

However you decide to approach such a project, whatever your desired end result, allow yourself plenty of time and consideration for how all the parts will fit together and function.

Chapter 4

Motherboard Upgrade

Now that we've done some 'body work' in Chapters 1 and 3, one of the three simplest upgrades you can perform on an existing PC is to give it a new heart, soul and engine – that is, replace the motherboard and CPU. The other two upgrades – increasing RAM and replacing an existing hard drive with a bigger, faster model, may come along as part of this project. I go through the process of motherboard upgrades as frequently as once a year for my main PC, and trickle the upgrades down to other test PCs and servers over a 1–2 year time span.

The advantage of replacing the motherboard, even if other components must be changed to suit, over buying an entirely new PC, is still economics. The biggest leap is switching to a motherboard with higher-power ATX power supply requirements, so a new power supply almost always goes into the mix at some point. If your motherboard is over three-years old you will probably have to upgrade the memory as well. For all of this potential cost – $150–250 for a new motherboard, $25–50 for a new power supply, and $100 for new memory – you get an upgrade for less than the cost of a new PC and your basic configuration stays the same. If you retain your current disk drive you have no hassles transferring data, re-installing programs and rearranging peripherals, but, changing the disk drive does give you a fresh start, more space, and more speed.

This project will take us through dis-assembly of an existing PC, replacement with upgraded parts, and enjoying the performance enhancements of a new life for an old PC.

TIP: Backup your data! Whenever you perform major work on your PC it is always a good idea, a life-saver even, to make a backup copy of at least your data files – documents, spreadsheets, photos, music, and such.

Tool Kit

- Basic assortment of flat and cross-point blade screwdrivers.
- Flashlight – to see into those dark nooks and crannies.
- Needle-nosed pliers to get things out of the nooks and crannies.

Parts List

- New motherboard and CPU of choice – current motherboards supporting and including the Pentium D dual-core processor can be found for as little as $149.

- New power supply – if your present system contains an older AT- or ATX-style system board it may not have the appropriate/additional power leads for the CPU and PCIe video required by the new motherboard.

- Upgraded RAM if your present motherboard does not use PC3200 or better-grade RAM modules.

Steps

Upgrading a motherboard is not a complex task – most of the connections are fairly obvious or well-documented in the system manual and the manual for the new motherboard. Patience is a virtue, however, as there are numerous connections, as you can see in Chapter 3, and getting to all the connectors and just removing and installing a motherboard can require significant disassembly of the PC chassis.

You can expect that your operating system (typically Windows) will discover new devices and require new device drivers when you start the new motherboard for the first time so keep that new driver CD handy!

TIP: If your new motherboard has a Serial-ATA (SATA) interface for your hard drives, install the drivers for the new board while the old is still working; this way the completed upgrade will be ready to boot properly and support the new SATA interface without problems.

Disassembly of the existing PC chassis for easy access to the motherboard, is where we'll start, then reverse the process putting things back together.

1. Disconnect everything from your PC – power cord, mouse, keyboard, LAN, phone line, printer, camera – everything. Start with the power cord first, Figure 4-1.

2. Remove the cover(s) from the chassis.

TIP: This is the perfect time to look inside the chassis and compare the current motherboard and available space with the new motherboard, including CPU, heatsink, and fan. I have

Figure 4-1   *Safety First! Unplug the power cord.*

frequently encountered chassis that will not accept a new motherboard because a larger CPU, heatsink, and fan will not fit behind the power supply – causing me to select a new chassis so everything will fit properly.

3. Remove the retaining screws from the brackets of any plug-in cards – video, LAN, modem, sound, etc. – then remove each of the cards.

4. Determine which if any components – disk drive, diskette drive, CD-ROM drive, or power supply – need to be removed to allow easy access to connections on the motherboard. Disconnect the power and data cables from each of the above then remove them.

5. Disconnect the power, data, and any chassis fan power cables from the motherboard, as shown in Figure 4-2. This should leave you with a clear field to work around.

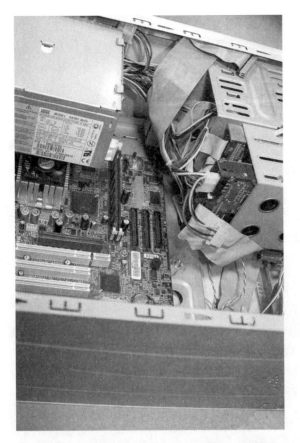

Figure 4-2 *Power and data cables disconnected from the old motherboard.*

Figure 4-4 *Removing the old motherboard from the mounting plate.*

6. Slide the motherboard and its mounting tray out of the chassis, Figure 4-3.

7. Remove the motherboard retaining screws and then the motherboard from the plate, Figure 4-4.

8. Place the new motherboard on the mounting plate to check for the placement of the mounting studs (screw holes), as shown in Figure 4-5. Check to see if you need to add or remove extra screw mounts.

TIP: Every hole in the motherboard should match up with a mounting point – there should be no

Figure 4-3 *Removing the mounting plate from the PC chassis.*

Figure 4-5 *Align the new motherboard with the mounting holes.*

Figure 4-6 *A nylon stud supports the motherboard where no screw hole is available.*

Figure 4-7 *Cut off the keystone button to keep the motherboard level with the other mounting holes.*

empty holes. Every mounting point on the panel should have a hole in the motherboard – there should be no extra mounting points.

9. If there is a mounting stud without an associated screw hole in the motherboard, the motherboard will be unsupported possibly causing damage. To support the motherboard where there is no mounting point, create one using a nylon mounting stud, Figure 4-6, cutting off the keystone button, Figure 4-7, if necessary to keep the board level.

10. After lining up the mounting holes, install the new bezel for your motherboard, Figure 4-8, in the rear panel of the chassis.

11. With the mounting holes and points lined up, secure the motherboard to the panel. Once the motherboard is secure, make sure any plug-in cards – video, LAN, etc. – fit properly into their respective sockets and mate with mounting holes for the brackets, then secure them with screws.

12. If not already done, install the CPU, heatsink and fan.

TIP: Follow the instructions carefully for applying heatsink compound between the CPU and heatsink. Heatsink compound or the pre-installed thermal pad exists merely to fill minute irregularities in the metal surfaces of the heatsink material and cap of the CPU chip – it is NOT a

glue, paste, caulk, messy coating, or crack-filler. Excess compound will reduce the ability of the heatsink to capture and remove heat from the CPU. Too little compound will similarly reduce heat removal capabilities and can trap moisture in the gaps which can boil and expand causing damage to the CPU.

Figure 4-8 *Connector bezel mounted on rear panel before securing the motherboard.*

13. Install the RAM modules in the new motherboard. If you had not anticipated it yet – this is one of the three most common, useful, and often necessary, upgrades I mentioned at the start of the chapter. If your previous motherboard and CPU were Pentium III or earlier vintage you will have to upgrade the RAM to accommodate the newest Pentium or AMD processors.

14. At this point you are ready for the first test of the new 'guts' of your PC – applying power and verifying that the board actually works. Chances are, if you pay careful attention to the mounting details and CPU installation, your system will work properly at this stage.

NOTE: You may test the motherboard before going through all of the work of disconnection and disassembly – setting up the motherboard and power supply on a flat surface with enough space to work on them.

- Connect the front-panel LED, power switch and other leads to the motherboard at the appropriate points, as shown in Figure 4-9. For this step you will have to refer to the instruction manual for the motherboard to know where these connections are.

- Connect your video display to the built-in or plug-in video adapter connection. Turn the display on so it is ready to present you with the Power On Self Test boot screen and setup instructions.

- Connect a keyboard to the new motherboard.

- Connect the power supply to the motherboard.

- Connect the AC power cord to the power supply, take a deep breath, cross your fingers, and then …

- Press the power button. Observe the CPU fan to make sure it begins to spin and stays spinning. If it does not stay spinning it is possible that something is overloading the power supply or one of the power

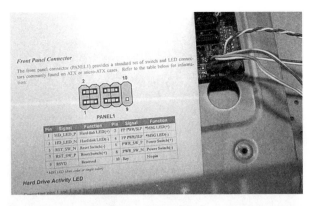

Figure 4-9 *Connections for the front panel switches and LED indicators.*

connections is wrong – *disconnect the AC power cord from the power supply immediately!*

Make sure you connected the CPU fan connector to the motherboard – many boards will shut down if the CPU fan is not detected. Some will even shutdown if the chassis fan is not connected – if you do not have one you will have to acquire and install one.

Check your connections and make sure the power supply has sufficient wattage rating to power the motherboard – a 200–250 watt supply may not be sufficient for the high-speed AMD and Intel CPUs.

- Within a few seconds you should observe the boot up screen from the motherboard BIOS and probably warnings that no hard drive or boot device was found. This is expected and perfectly OK.

- Turn off the system and disconnect the AC power cord. It is time to finish the assembly work.

15. Reinstall any disk drives removed in Step 4 above. Reconnect them with new data cables supplied with the motherboard and hook up the power connectors, Figure 4-10. This should complete the basic reassembly and allow your system to boot into Windows.

Figure 4-10   *Disk drives connected to the new motherboard and power cables.*

16. When you turn on the power to the system with the disk drives connected the system BIOS may display new errors or status messages with respect to the 'new' devices the BIOS encountered and ask you to adjust the settings in the SETUP program, or merely acknowledge the new devices and allow you to continue to boot up.

17. Place the driver CD for your new motherboard in the CD-ROM drive. Upon booting into Windows you will probably encounter successive *Found New Hardware* messages and wizards prompting you to install drivers for the new motherboard chipsets and components. Point the wizard process to the new driver CD to pickup necessary drivers, and restart the system when prompted so things load properly.

18. When all new device drivers are loaded and the system is restarted most everything should function as you expected, only faster! Beware that this is a significant change to the overall system Windows used to know and you may see messages about reactivating Windows. Scary as this may seem, don't fret, you've done nothing wrong and will not be admonished – simply call the 800 # provided, explain to the support technician that you've upgraded your motherboard, follow the instructions you're given over the phone, get a new activation code and reactivate Windows.

19. Shutdown the system, disconnect the AC power cord, finish reassembling the chassis parts and covers and call it done. Congratulations!

Summary

Upgrading a PC motherboard is one of the most exhilarating projects you can undertake – full of anticipation, complexity, and best of all – results! If you combine this project with any of the three previous projects you'll know most of the ins-and-outs of a PC and what makes it tick. Having a faster, newer PC is as beneficial as it is exciting. Doing fun and interesting things with a great PC is yet to come.

Chapter 5

Build Your Own Power Protection

What's your PC plugged into? A normal outlet strip offers no surge protection at all – it's just an extension cord with 4–6 outlets at the end. A surge-protection outlet strip provides some protection – perhaps enough to save your PC from only the most common electrical power surges generated by nearby appliances.

The typical surge-protection outlet strip differs from an extension cord or plain outlet strip by the addition of one small part – a metal-oxide varistor, also known as an 'MOV'. This marvelous quarter-size device acts like a voltage clamp or limiter to keep high voltages (power surges) away from your PC. When the AC line voltage reaches 130 or 150 volts the MOV effectively 'shorts-out' the power source causing a protective fuse or circuit breaker to open the circuit and remove power from the attached devices – your PC, printer, display, etc. This is crude but highly effective in some cases.

An MOV can take only so much high voltage abuse before damaging surges destroy it, leaving you with no protection at all. You cannot depend on just a single, simple surge protector to defend your electronic devices against really big power surges such as those that accompany lightning storms or just plain lousy utility power (which may be common in older buildings or rural areas). Bad power's effect on your PC and other electronic items may not be immediate or obvious. Power surges stress all of the components in the power supply and they can fail over time – eventually causing intermittent or complete failure. Since you've made an investment in your PC, and probably plan on investing more in souping it up, why let it succumb to bad power?

The simple parts in most household electrical plugs, outlets, and extension cords cannot withstand the super-high voltages of a direct or nearby lightning strike – so do not expect any power protection device to save your gear under any circumstances – but you can take measures to provide cleaner electrical power to your stuff.

'Cleaner electrical power' means AC power with little or no noise, static affected by minor power surges. If you've ever experienced the interference an old mixer or vacuum cleaner motor could inflict on a television picture or AM radio signal you have a pretty good idea what I am referring to. Interference, noise, and static – all about the same thing – have no business on the power line or feeding your precious PC.

If your PC acts a little funny when the dishwasher, clothes dryer, refrigerator, or vacuum cleaner turn on or off – you probably have a power problem which may affect your PC. Unfortunately you've probably had no idea how to get rid of it – until now. If the lights in your computer room dim or your monitor flickers when heavy appliances are in use, then you have low power which can be helped by the use of an uninterruptible power supply (or potentially dangerous bad wiring, for which you need an electrician and serious repairs.)

This project is about building a power line noise filter and protection device you may only be able to match by spending $60 or more on a truly sophisticated power protection device. For less than $30, plus or minus some scrounging, you'll get some new experience and peace of mind.

The core components of this project are not one but three MOV parts for additional surge-protection, and a special power-line filter that reduces noise and will help soften some power surges.

CAUTION! You will not only be using power tools (a drill) and soldering iron (very hot) but you will be building an electrical device that will connect to potentially lethal 120 volt power lines. If you are not comfortable with electrical items and safety, please stop here.

Parents or an advisor competent in electrical circuits should supervise the building and testing of this project.

Tool List

- Drill motor
- Drill bits – 5/32- and 1/4-inch drill bits
- 1/2-inch and 5/8-inch Greenlee or equivalent chassis punch
- Soldering iron
- Solder for electronics (60/40 or 63/37 mix, not 50/50 or no-lead plumbing solder)

- Wire cutters
- Needle-nosed pliers
- Flat-head and cross-point screwdriver
- Volt-ohm-meter to safely test your circuit
- Safety glasses/goggles.

Parts List

- 6 foot long 3-wire 'grounded' 12- or 14-gauge extension cord
- 2×3×6 inch metal chassis (Radio Shack catalog number 270-238)
- Corcom 10VR1 (Mouser.com part number 592-10VR1, or Kobiconn 10CA1, Mouser.com part number 437-10CA1) or equivalent dual-stage AC line filter
- 3 each Littelfuse TMOV14R130M (Mouser.com 576-TMOV14R130M) or Radio Shack 276-568 130 volt MOV components

- Fuseholder for 3AG-size fuses (Mouser part, RS part)
- 10 amp 3AG-size fast-acting fuses
- Grommets – to insulate and protect the power cables, or 'Hyco' cable bushings
- Wire tie 'zip' strips to retain the power cords in place (if using grommets)
- 12–18 inches of solid or stranded hookup wire (may be taken from within the power cord)
- 6–32 × 3/8–1/2 inch long machine screws and matching nuts to mount filter to chassis

The Schematic

There are three significant parts of this project, shown schematically in Figure 5-1 – a fuse, to provide the fastest circuit-opening protection available; the power filter 'brick' that reduces interference and offers some

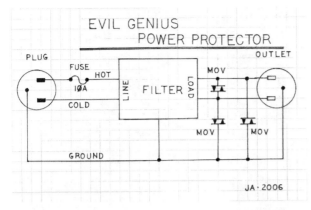

Figure 5-1   *Schematic of the power filter and protection project.*

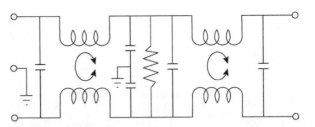

Figure 5-2   *The internal schematic of a typical power filter network.*

buffering to power surges so the MOVs don't go into action and blow the fuse; and the MOV surge protectors, closest to the output side, to keep voltage spikes to a minimum.

Power comes into the circuit through a standard power cord, connects to the fuse as a first-line of defense. The filter and MOVs do the work and clean, protected power leaves through the outlet cord to the load – your PC and such.

Using a fast-acting rather than 'slow blow' fuse is a 'must' in a protection circuit like this – the MOV is pretty slow reacting to surges but when it reaches its limit and clamps down on spikes you want the circuit to open up and power removed as quickly as possible.

The filter, Figure 5-2, is designed to significantly reduce noise, radio frequency, and some fast voltage spike interference. This keeps a lot of minor interference from getting to the MOVs leaving them do only the 'heavy lifting' of shorting out the most significant surges.

Figure 5-3   *Components of the power protection project.*

Power surges come along the wires in odd ways. Using three MOVs offers protection against power surges that may occur between any three wires of the power line: ground and the HOT side of the AC power line; ground and the COLD/Neutral side of the AC power line; and the HOT and COLD lines.

The components, Figure 5-3, will handle a 10 ampere (1200 watt) load – enough capacity to run a PC, CRT monitor, ink-jet style printer, DSL or cable modem, and a home router. If you need to protect more devices you will need to build more units. Overall this is a very simple circuit with a very simple purpose – to provide cleaner power to your PC and peripherals than a plain-old outlet strip.

Steps

1. Lay the components out in the metal chassis, Figure 5-4, allowing working room and a safe distance of 1/4–1/2-inch of clearance between all electrical connections and the sides of the metal chassis.

2. Mark the places to drill the 5/32-inch holes for mounting the power filter and 1/4-inch pilot hole for the chassis punches to make the holes for the cables and fuseholder.

Figure 5-4   *Components laid out in the chassis.*

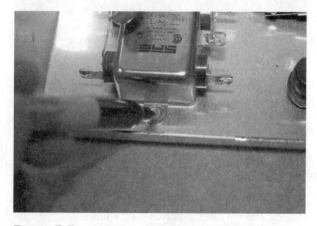

Figure 5-5   *Marking the holes for the filter.*

Figure 5-6   *Marking the holes for the fuseholder and power cord.*

3. Drill the 5/32-inch holes for the power filter and the 1/4-inch pilot hole, Figure 5-7, for the chassis punch.

4. Using the 5/8-inch chassis punch, Figure 5-8, complete the holes for the fuseholder and Hyco

Figure 5-7   *Drilling the pilot hole for the chassis punch.*

bushings (use the 1/2-inch punch if using grommets and wire ties to secure the cord).

5. Mount the power filter and fuseholder in the chassis, Figure 5–10. Observe the power cord secured with a Hyco bushing.

6. Connect the leads of one MOV across the LOAD terminals of the power filter (MOV1 in the schematic), trim excess wire but do not solder yet.

7. Connect an MOV from one of the power filter LOAD terminals to the Ground terminal, then repeat for the other terminal. Trim excess wire, but do not solder yet.

8. Cut the extension cord into two pieces. At your discretion the portion of the cord with the plug can be longer, shorter or the same length as the portion with the socket. (You may have already done this to snag that couple of inches of hook-up wire for the fuseholder to filter connection.)

Figure 5-8   *Using the chassis punch to make 5/8-inch holes.*

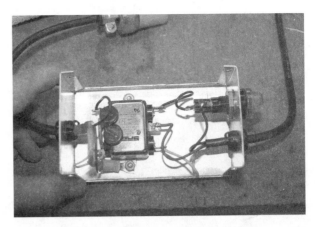

Figure 5-10   *The completely wired power filter.*

Figure 5-9   *The main components in place ready for wiring and soldering.*

9. Carefully slice and remove 3–4 inches of the outer insulation jacket from each power cord.

10. Insert the cord with the outlet in the grommet nearest the LOAD terminals of the power filter.

11. Insert the cord with the plug in the grommet nearest the fuseholder and LINE terminal of the power filter.

12. Measure, cut, connect and solder a wire from one end of the fuseholder to one of the input terminals at the LINE side of the power filter.

13. Strip off 3/8–1/2-inch of insulation from each of the wires.

14. Twist the conductors of each wire tightly to keep them from fraying.

15. Heat the wires with the soldering iron then apply a small amount of solder, allowing it to melt and flow into the twist of wires – this is called 'tinning' the wires to make it easier to solder them on the terminals.

16. Connect then solder the green wires from both power cords to a Ground terminal of the power filter. This should secure two of the MOV leads, the two cords, and the filter/ground together.

17. Connect then solder the black wire from the cord with the outlet to the LOAD terminal directly opposite the LINE terminal connected to the fuseholder. This should secure two of the MOV leads, the filter and the cord together.

18. Connect then solder the white wire from the cord with the outlet to the LOAD terminal directly opposite the empty LINE terminal. This should secure the remaining two MOV leads, the filter and the cord together.

19. Connect then solder the black wire from the cord with the plug to the empty terminal of the fuseholder.

20. Connect then solder the white wire from the cord with the plug to the empty terminal on the LINE side of the power filter. This completes the wiring and connections, as shown in Figure 5-3.

Figure 5-11   *The completed power protection project.*

21. If you used grommets instead of Hyco bushings, secure both power cords with wire 'zip' ties so they cannot be pulled out of the chassis.

22. Install one of the 10 ampere fuses in the fuseholder.

23. If you have a volt-ohm meter, now is a good time to test the circuit for improper shorts or open connections.

There should never be any continuity (should be infinite resistance) between any of the black wires and either white or green wires.

There should never be any continuity (should be infinite resistance) between any of the white wires to either the black or green wires.

There MUST be complete continuity (low/no resistance) from the black wire at the plug to the black wire at the outlet, and the same respectively for white-to-white and green-to-green, so the project can deliver power to the loads.

Provided you have no shorts between the Hot/Black and Cold/White or Ground/Green wires your new circuit should be safe and ready for a 'hot test' with full AC line voltage. If you have no continuity from input (Line) to output (Load) side of the circuit (Black to Black, White to White and Green to Green) nothing will work, but that indicates one or more bad components rather than a dire safety hazard.

24. Assemble the cover to your chassis, Figure 5-11, watching closely to be sure none of the internal connections make contact with the cover. Repeat Step 22 to verify the integrity of your work.

TIP: The next steps involve connecting your project to live AC power. This is a good time to make sure you are comfortable and sure of your work – and get out those safety glasses!

25. Plug your completed power protection project into one of your other outlets strips – preferably one with an on/off switch (in the OFF position) and a fuse or circuit breaker for protection.

26. Step back a bit from the project then turn the outlet strip power switch ON. If all is well nothing will happen – no sparks, pops, smoke, flames – nothing dramatic would be great news!

27. Turn the outlet strip OFF.

TIP: If you did encounter something bad there is an undetected problem with your circuit and you need to double check the wiring, replace the fuse(s) and try again.

28. Plug a simple table or work lamp into the outlet of your power protection project (check to make sure it is turned on and works first, leave it on) then turn on the switch on the outlet strip – the lamp should light!

29. Congratulations! This may not seem like a very exciting accomplishment but in reality it proves you can safely and successfully make a device that handles high voltage and that delivers power to a load.

30. Disconnect everything, unplug the outlet strip from the wall and plug it into the outlet of your project.

31. Plug your new power protection project into the wall socket, then connect your PC and peripherals to the outlet strip – they can now run with cleaner power and be better protected!

Summary

Tossing a handful of components and wires into a box and plugging it into the wall is a fairly significant undertaking. Protecting your PC and peripherals from the threats of bad power and electrical surges is no trivial matter. The best factory-made power protection products are priced well over $60 and you've created an equivalent for less than half the cost. With cleaner power the rest of your PC projects should run smoothly.

Chapter 6

Secure Your WiFi

Most of us have and really enjoy wireless networks. They allow tremendous flexibility and convenience. What you may not know is that others may also be enjoying your wireless network – uninvited and unknown to you – because your current WiFi system may not be protected at all, or the protection it does have uses the potentially hackable Wired Equivalency Protocol (WEP).

Why is security important? Remember, WiFi is wireless, radio waves, typically broadcast in all directions just like your local TV and radio stations. WiFi is essentially unbounded compared with a wired network whose signals do not go outside the cable and jacks connecting computers to hubs and routers. WiFi radio waves are a broadcast of data that can be picked up hundreds if not thousands of feet away under the right conditions. WEP was the first type of security offered for WiFi – an attempt to keep the radio waves almost as secure as if you were using wires – hence the name Wired Equivalency Protocol.

Unfortunately WEP is nowhere near as secure as a wired network. Either 'strength' of WEP, 64- or 128-bit encryption, can be 'cracked' in a matter of minutes with software written specifically to sample the wireless data stream and continuously try sequences of codes until the resulting data output looks like TCP/IP packets. When this 'cracking' is accomplished your wireless network is usable by anyone with 'the code', to send or receive data – putting your data and your Internet connection at risk.

What are the risks? The first risk is that someone else will be stealing the Internet access you pay for – and letting people outside your home use your Internet connection may be grounds for canceling your service. The next risk is that someone unknown has access to the computers in your home – putting your personal data at risk. If the first two aren't bad enough, your Internet connection may be abused to send SPAM, pornography, viruses, or hacking attempts over the Internet, making it look like you are the guilty party.

Securing your wireless network can avoid myriad legal and personal complications, and provides an opportunity to kick up your wireless performance another notch or two by replacing your older, less secure, slower 802.11b access point with a new 802.11g access point that provides at least five-times the speed and stronger Wireless Protected Access (WPA) security.

Since WPA hides the data encryption codes on and between your computer and the wireless access point, it is significantly more difficult to determine the secret codes and infiltrate the network connection.

Upgrading to 802.11g or the upcoming 802.11n-MIMO standard (due to be more available sometime in late 2006 or 2007) can boost your WiFi LAN throughput from a typical 5–10 megabits-per-second (Mbps) to 54 Mbps or more, to nearly match the most common corporate networks of 100 Mbps. Since most of us are still surfing the web over cable or DSL Internet connections of 1.2–7 Mbps we will not see an improvement in web performance, but many current and intended home networking applications such as music and video sharing over WiFi will benefit significantly with higher speeds.

NOTE: The MIMO (multiple-input/multiple-output) based 802.11n network products have not been standardized or made widely available from all WiFi product vendors because the IEEE 802.11n standard is not in final,

Figure 6-1   *A variety of older 11 Mbps, non-WPA 802. 11b WiFi products.*

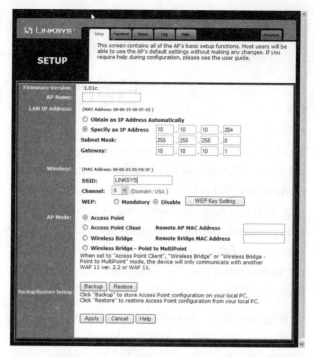

Figure 6-2   *The configuration page for older products show that WPA is not one of the security options.*

production-ready form. Some vendors, most notably Belkin, are selling Pre-N WiFi products. While it is nice to play on the bleeding edge of technology, spending money on 'unofficial' products could leave you with incompatible products in the future.

The first step to determine if you need either a security or speed upgrade is to verify the specifications of your existing WiFi gear. Older Belkin, AirLink, Linksys, D-Link or Netgear products, such as those pictured in Figure 6-1, that do not specifically state 802.11g or 'G' in their literature or on the product labeling, or state only 802.11b, are candidates for an upgrade to 'G'.

Speed is nice but security is our biggest concern and you want to make sure your WiFi gear and your PC will support WPA security. Support for WPA is not widely advertised or marked on the products. You have to consult the documentation or support website for each vendor and check the model number and specifications for every device you have to determine WPA-support. Also look to see if there is a WPA-supporting firmware or driver

upgrade for your WiFi gear. You can even look at the configuration web page built into your WiFi access point, Figure 6-2, to see if it offers WPA as one of the security options.

Another thing to note about the settings in Figure 6-2 is the SSID has been left at the default of 'Linksys' – a significant clue to hackers that you have probably not configured your access point to be more secure.

As for your PC, WPA is supported only in Windows XP Home or Professional with Service Pack 1 or 2 installed. WPA is not available for Windows 95, 98, Me, 2000, or original XP. If you haven't upgraded to XP Service pack 2 yet, now would be a good time!

When selecting new WiFi hardware be diligent about checking all of the specifications, from the type of 802.11g support provided to WPA. If you care to dig deep, also research the chipsets used in various products or comparison-shop interbrand compatibility. An excellent overview is provided in a PC World article explaining 802.11g performance: http://www.pcworld.com/reviews/article/0,aid,116279,00.asp.

You will find numerous brand names and technology claims that have nothing to do with the official 54 Mbps 802.11g standard – 'Super G', 'Extreme G', 'Turbo G', 'Afterburner', etc. – each referring to a particular brand or chipset-specific mode of implementing higher speed '802.11g'. In reality, makers D-Link and Netgear use the Atheros chip claiming 108 Mbps throughout (which is only theoretical) while LinkSys and Belkin are claiming 125 Mbps performance using a Broadcom chipset.

All of these products should be compatible at the base 54 Mbps 802.11g performance levels, but you will not get 108 or 125 Mbps mixing LinkSys or Belkin with D-Link or NetGear products – they just do not work the same way above 54 Mbps.

Another caution is that the 'Super G' devices, operating in their respective 'Super G' modes will use two of the available three nonoverlapping WiFi channels – meaning you cannot use these products in a multi-access-point (AP) network without some problems. I have two access points, one to cover inside my home, the other to cover my office and patio areas, in which 'Super G' does not work in the overlapping coverage areas between the two APs.

The 'Afterburner' products stick with one channel and cram more data into it, making it a more neighborly choice for your own and nearby WiFi networks. (I've tested the USRobotics products before and find them to be superb when using all and only USR equipment, but I have a mix of Intel, Atheros, Broadcom and other WiFi devices to accommodate as well.)

Your next consideration is whether to get a WiFi access point with built-in router/firewall or a simple access-point without router/firewall. If all you have

Figure 6-3 *A variety of new 54 Mbps, WPA-compatible 802.11g WiFi products.*

is a DSL or cable modem without a separate firewall/router, the choice may be somewhat obvious – get the full-featured access point.

I prefer to keep my networking functions separate and more flexible, using a separate modem, separate router/firewall, then adding an access-point-only for WiFi connections. This lets me easily shutdown or reconfigure the wireless network without affecting the systems and connections on the wired-LAN. With all this information in mind, let's walk-through a typical wireless upgrade.

Parts List

- 802.11g access point with WPA support – in this case a D-Link DWL-G700AP.

- 802.11g wireless cards for laptops – for this project I happen to have a

D-Link DWL-G630 card, so I can expect compatibility between devices.

1. Disconnect your existing WiFi access point.

2. Connect your new WiFi access point – both power and network connection. You cannot configure your access point over wireless – as this would provide a serious security risk. You must use a wired network connection between your PC and the access point (and this connection may be through your network hubs/switches) to access the built-in setup web pages for the access point.

3. Following the typical instructions for your access point, set your browser to go to IP address of the access point – for our DWL-G700AP the address to use is 192.168.0.50.

4. When the access point answers the browser request you will be prompted for a username and password to log into the access point – the default User name is 'admin', there is no password, so type in 'admin' then click the **OK** button.

5. In the middle of the access point's 'home page' is a 'Run Wizard' button. Click it to begin the configuration of the AP, Figure 6-4, then click the **Next** button to go to the next dialog, Figure 6-5 to set the admin password

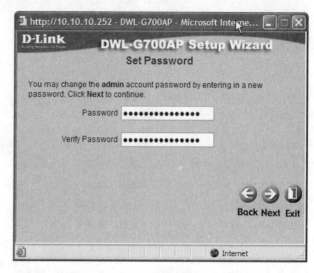

Figure 6-5 *Setting a local administrator password for your access point.*

for the access point, then click the **Next** button.

6. In the next dialog, Figure 6-6, change the SSID (name) for your wireless network to be something you recognize, but not something that identifies you or your location, just to obscure that it is your wireless access point, then select a Channel. Click on the **Next** button to go to the last setup dialog.

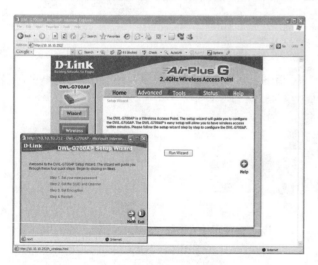

Figure 6-4 *The start of the setup wizard for a D-Link access point.*

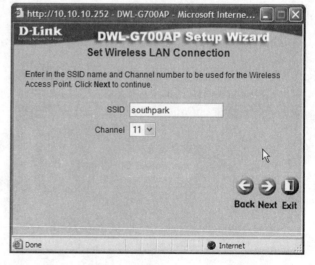

Figure 6-6 *Renaming the access point SSID and selecting a channel.*

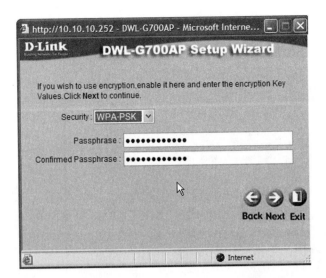

Figure 6-7 *Renaming the access point SSID and selecting a channel .*

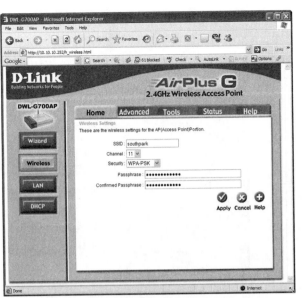

Figure 6-8 *Renaming the SSID, selecting a channel and setting security values for the access point.*

TIP: While 802.11b provides eleven possible WiFi channels, most of them overlap adjacent channel numbers and cause interference, poor performance or keep your WiFi network from working at all. Of the eleven available channels only three of them do not overlap – Channels 1, 6, and 11. When you have two or more access points nearby, on the same network or not, you need to choose non-overlapping channels. Most access points are set to Channel 1 or 6 'out of the box' so you can expect the most interference and obvious hacking attempts when using either of these channels.

7. The final setup dialog, Figure 6-7, allows you to select the type of security and encryption you will use to secure your WiFi network, as well as provide the passphrase. Windows XP

supports both WEP and WPA-PSK. WPA is more secure and allows greater flexibility/more combinations of the 'secret code' each wireless client must 'know' or be set for to connect. Choose WPA or WPA-PSK and then type in your secure phrase – combinations of letters and numbers, or a complex phrase, are more secure. Click on the **Next** button to complete setup.

You may also access these settings and more by using the tabbed dialogs and menu selections from the access point's main control screen, as shown in Figure 6-8. When you have finished setting the values you want to use, restart the access point to activate the settings.

Steps to Install a New WiFi Card

1. Remove your old wireless card(s) from your PC(s). Install the drivers for your new wireless card on your laptop before connecting the new WiFi card to the PC.

2. Connect/insert your new wireless card and follow the procedures to ensure the new

drivers are installed and working correctly. If all is well you should be alerted to the system discovering a new wireless network, providing you the option to connect/configure your system to it. The next few steps will show you how to setup the more secure connection used by your new access point.

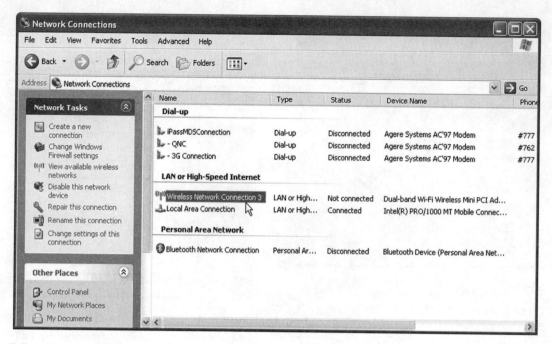

Figure 6-9 *Windows Network Connections shows available adapters.*

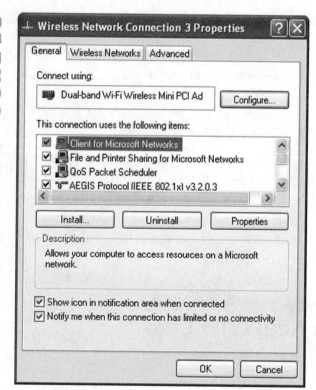

Figure 6-10 *Windows Wireless Network Connection Properties.*

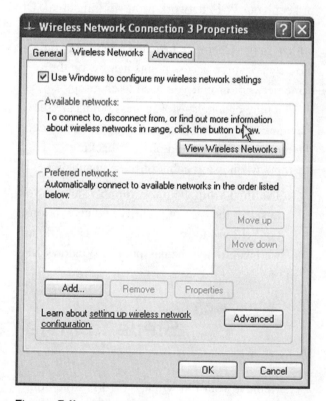

Figure 6-11 *The Wireless Network dialog reveals access to all wireless settings.*

3. If the Wireless Network Connection Properties dialog, Figure 6.10, does not appear, click **Start**, select **Control Panel** then

double-click **Network Connections** to reveal the available network adapters, Figure 6-9. Your new wireless adapter should be among

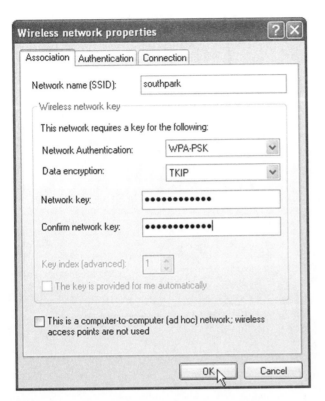

Figure 6-12 *Windows Wireless Network Connection Properties provides access to security settings.*

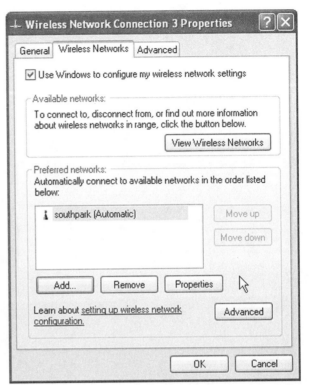

Figure 6-13 *Available networks appear in the Preferred networks list.*

them. Right-click the new adapter then select **Properties**. Select the **Wireless Networks tab**.

4. On the dialog that appears under the Wireless Networks tab, Figure 6-11. Click the **Add** button to access the settings you need to adjust for your new network, Figure 6-12.

5. Enter the SSID name you used in configuring the access point. Select **WPA-PSK** for

Network Authentication and **TKIP** for Data encryption. Then type in the key or passphrase you used when setting up the access point. Click the **OK** button when all settings are entered, you'll see your new access point in the preferred networks list, Figure 6-13.

6. Click the **OK** button to close the properties dialog. Your wireless network settings are 'good to go'.

Summary

Installing or replacing a wireless network is typically an unpack, plug, and play process but playing with only the default settings leaves your network vulnerable. Knowing and using the proper settings can ensure reliabilty, best performance and proper security for your network and data.

Chapter 7

Dynamic Picture Frame

You've probably seen digital picture frames at specialty stores and online retailers and thought them to be a way to get Grandma into the digital age by sending her an entire photo album in an electronic 'frame', but otherwise just a cute but expensive novelty. At $150 these gift items are truly novel uses of LCD displays and a bit of computer guts, but you can create your own dynamic picture frame and let it display updated, perhaps even seasonal pictures at your whim!

Well, that seems sort of obvious, after all, many of us have web-based Windows wallpaper and screen savers for our PCs so why not apply the same concept to a display piece in your den or as a gift item?

There isn't much keeping you from mastering this project beyond an inexpensive laptop, some 'slide show' software, and your choice of pictures. You may not have a laptop for regular use around the house and here I am suggesting you procure an inexpensive laptop just to display pictures. Well, you were going to spend $150 or more on that very novelty item for Grandma, why not the same on yourself?

You really do not need a powerful PC, or even one with a display resolution greater than 800×600 or 1024×768. You can use Windows 95, 98, Me or XP – your choice – to run most slide-show display software. For storage a mere 8–12 GB hard drive will do quite well, and 64–256 MB of RAM is plenty to get an operating system started and display your digital images.

OK, so, what model laptop to buy? Most inexpensive and readily available Toshiba, Hewlett-Packard/Compaq, and Dell PCs are probably too bulky for the sofa table or curio shelf. One obvious choice is probably an IBM ThinkPad 600, 600E or 600X – somewhat thin, lightweight, and certainly capable of running any slideshow software to show off the family. Besides being relatively thin, the ThinkPad-series of laptops are finished in flat-black and generally show-off pretty well in many a décor. Better if you can find a T20, T21, T22, or T23, but then you just might want to use one of those for real computing.

These older models can be found on eBay and at surplus/refurbished computer centers for around $120. Add a wired or wireless network adapter and you will be able to change the slideshow through your home network by remote control.

Figure 7-1   *An ordinary laptop waiting to become a wired or wireless connected piece of art.*

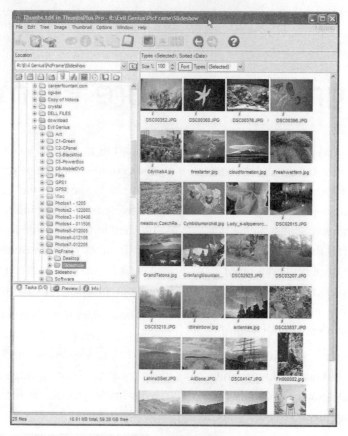

Figure 7-2 *Using ThumbsPlus to preview and select the pictures for a slideshow.*

A laptop by itself may not be a very attractive piece for your living room with the keyboard hanging out, so you could look for a more expensive tablet PC you can fold over to show just the display, or, hide the keyboard and display within a custom-built wooden frame stained or painted to suit your décor. Since I don't often get to my shop to do woodworking because of all the technical projects I'm involved with, building a wooden frame for a laptop seems like good therapy, and a way to get off my butt and mix hobbies together in a mutually productive way.

No, this project won't be about woodworking and choosing the right stain and finish, but it may inspire you to create something special as we've done with the frog-green kid's PC or 'The Matrix'-black-techno-look PC makeovers.

Parts List

- Inexpensive laptop – $120 or so.
- Windows operating system – XP works great as it includes most of the drivers you will need plus it offers built-in remote control capability.

- 'Slide show' software – Google's Picasa, Microsoft Photo Story 3 for Windows, or my preference, Cerious Software's ThumbsPlus.
- Wired or wireless LAN card (optional).
- Megabytes of family or scenery photos.

Figure 7-3 *Selecting Image/Build Slideshow for selected pictures.*

Steps

1. Start with a clean, fresh Windows installation. You won't need to run but a couple of applications for this project so memory and disk space will not be too critical. If the PC is to be connected to your home network by wire or wireless, be sure to install virus and spyware protection, and desktop firewall software, or at least turn on Windows firewall. Figure 7.1 shows our plain-jane laptop with accessories, ready to be transformed.

2. Install the 'slide show' software of your choice – Google's Picasa, Microsoft's Photo Story 3 or Cerious Software's ThumbsPlus. (Yes, you could use Microsoft PowerPoint in slide show mode but that's a pretty expensive program to use for displaying pictures.)

TIP: Both Microsoft's Photo Story and Google's Picasa allow you to create single digital movie files, cycling through your selected digital photos with very smooth fades and zooms. Photo Story makes a Windows Media Video (WMV) file of your photos, and allows you to associate an audio file as the sound track for your movie, but you must play the slideshow using Windows Media Player. Picasa's slideshow feature creates an AVI video file, suitable for viewing with many different players, but does not allow you to include an audio track.

ThumbsPlus creates a slideshow as a stand-alone executable program that does require a separate player program, but does not allow you to include audio in the slideshow. Rather than stress over which file format and media player to use,

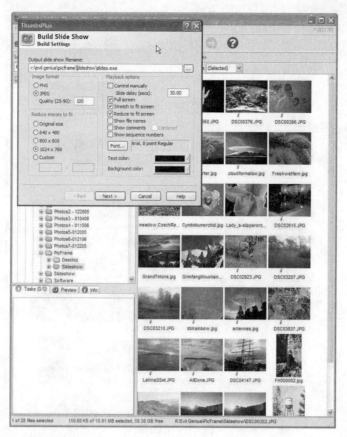

Figure 7-4   *Setting the ThumbsPlus slideshow filename and parameters.*

and because ThumbsPlus handles myriad other digital image chores for me (such as processing the photos I took for this book) I'll walk you through the steps of using it to create a slideshow.

3. Select the digital images you want to display in your show – they can be JPEG, GIF, BMP, TIFF, or just about any file format – so long as it is viewable in ThumbsPlus, and most are. It is best to copy your images into a single, easily identifiable folder so you can manage them as a group.

4. Run ThumbsPlus and navigate to the folder containing the pictures of interest – in my case \EvilGenius\PicFrame\Slideshow, as shown in Figure 7-2. From here navigation is easy and in three more steps the slideshow is ready to go.

5. Once you've selected the pictures you would like to display, start the slideshow building process by going to the menu bar, selecting

Image, then click on **Build Slide Show**, as shown in Figure 7-3.

6. In the Build Slideshow dialog you have to specify the name and location of the slideshow program you are creating, the format of the images, if you want to manually advance the show or have it change images automatically, how long you want each 'slide' to appear, and if you want details about the picture to appear as a caption.

As shown in Figure 7-4, I typically use JPEG files (as they come straight off my camera, or as shared by most people), prefer a high-resolution 1024 × 768 display format, automatic slide changes lasting 30 seconds each and no text on the screen. Click the **Next** button to proceed to the final step.

7. The last step in the slideshow creation process, Figure 7-5, is to decide whether or not to display a text message as an introduction to the slide show, which I prefer not to do, as it

Figure 7-5 *Adding an introductory message and creating the slideshow file.*

would only be seen once in our picture frame. Click the **Finish** button and the slideshow program is created for you.

In a few seconds your pictures are formatted and combined into a self-sufficient, stand-alone display program you can run on your picture frame laptop, and even send to your friends. Copy this program file into Windows' Start Menu, Programs, Startup file folder so your picture frame will run by itself.

The process is similar for Picasa and Photo Story, as shown in Figures 7-6 and 7-7.

The steps above reward you with the basics of a digital picture frame. There are three more things you can do to really enhance this project – add music to accompany your pictures, setup remote control network access so you can change your slideshow without disturbing the laptop or other nicknacks around it, and finally, adding a simple frame to hide the laptop and make the project more presentable.

Musical Picture Frame

We've had background music, mood music, elevator music, but your picture frames have probably been silent – until now – with the invention and release of new, perhaps even improved, 'picture frame music'. Yes, the very same laptop proudly displaying your family or vacation photos can not only bring back visual memories but auditory ones as well.

Make up a Windows MediaPlayer playlist of suitable background music, rushing waterfalls, chirping birds, or pounding surf to go with your picture theme and you'll have a truly dynamic, evocative picture frame that will entertain you and your guests for years to come.

When football season comes around run a slideshow of your favorite team plays and add their fight-song or play-by-play action. During the holidays change up the scene and add appropriate

Figure 7-6 *Building a slideshow with Google's Picasa.*

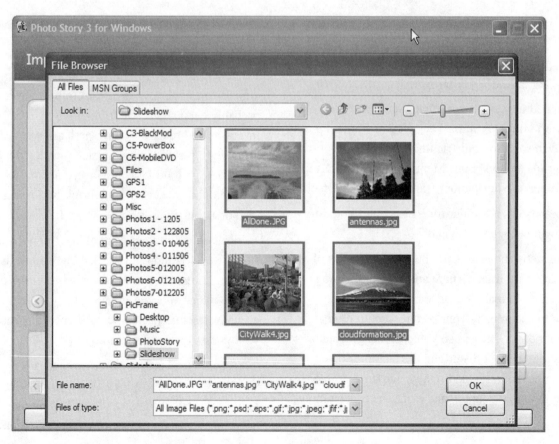

Figure 7-7 *Selecting files for a slideshow with Microsoft Photo Story.*

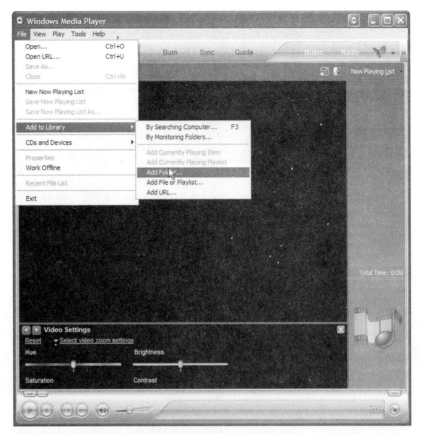

Figure 7-8   *Starting a new playlist with Media Player.*

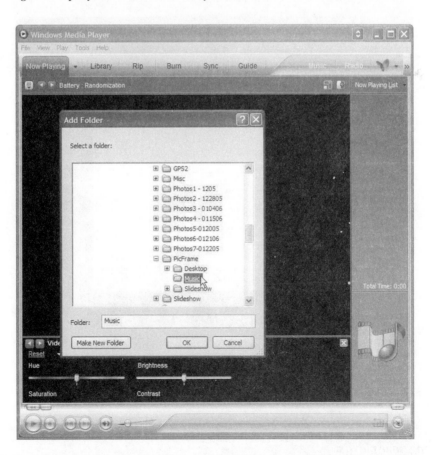

Figure 7-9   *Selecting the folder for Media Player to find files for your playlist.*

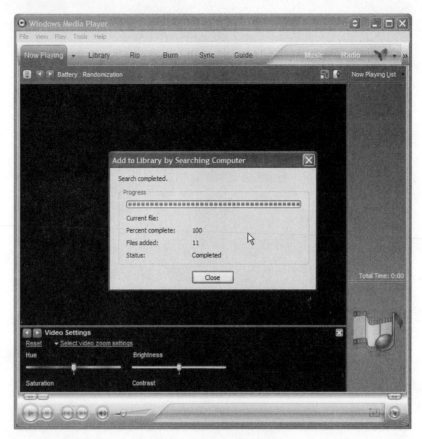

Figure 7-10 *Windows Media Player creating a playlist of songs.*

Figure 7-11 *Saving your playlist.*

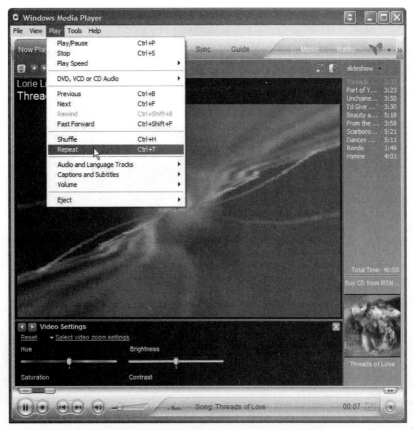

Figure 7-12 *Setting Windows Media Player to repeat a playlist of songs.*

music. Play a video of a waterfall with sound. This isn't some crazy gadget – it can be a trendy seasonal excuse to learn about multimedia and displace that portrait of grouchy old Aunt Gretchen!

To bring sound to your sights follow these steps:

1. Copy your favorite MP3 or WAV audio files to a folder on your picture frame laptop.

2. Open Windows Media Player, Figure 7-8, select the File menu, then Add to Library, then Add Folder…

3. Navigate to the folder containing your music files, Figure 7-9. Media Player will build a temporary playlist, Figure 7-10, which you can edit and then save, Figure 7-11.

Once your playlist is created and saved, when you play it you can indicate you want it to repeat, Figure 7-12, just like your slide show.

Remote Control

To change the contents of your slide show over your network, simply enable Windows XP's Remote Desktop (if you are using Windows XP Pro) or install one of many variations of the VNC remote console server program such as RealVNC or UltraVNC, on the picture frame laptop. VNC works equally well on Windows XP Home, Professional or Media Center editions.

Using a PC with the companion viewer/remote control program you can connect to and control the picture frame laptop over your network and do nearly everything you could do as if you were standing in front of it. Figure 7-13 shows a remote PC controlling the picture frame laptop using Windows Remote Desktop.

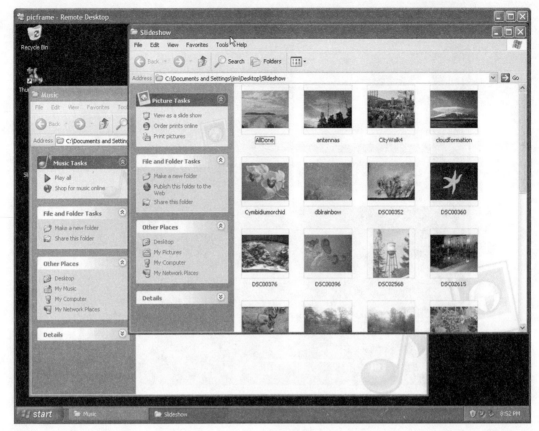

Figure 7-13 *A view of Windows Remote Desktop controlling our picture frame laptop.*

Frame the Frame

Once the electronics of your project are setup and running you may wish to consider a decorative touch to enhance the presence of your 'artwork' amidst your décor. For my application I've made an oak 'coverlet' to surround the main body of my laptop and cover the keyboard, making a space to set little nicknacks, and framed the laptop display so all you see is oak and a nice high-resolution digital picture.

The cover over the keyboard is simply three small boards on edge to make the sides – front, left, and right, and a sheet of veneer for the top – all glued together. I left clearance on the sides for some ventilation and space to let the display frame fit.

The display frame is narrow flat boards cut proportionately and glued with strips at the back. The sides extend down to and wrap around the sides of the base, but not glued or fastened in anyway, allowing the display to be tilted to an

Figure 7-14 *Completed picture frame laptop 'wrapped' in oak.*

appropriate viewing angle, and without altering the laptop at all. Apply some stain or finish to suit your taste or décor, add a few decorative or sentimental nicknacks and you are done.

Summary

You've probably realized that you will spend more time with the table saw, router, coping saw, finish nailer, wood glue, putty, sandpaper, and stain than with the laptop and picture files, but consider what you've created – a décor-friendly techno project that will evolve through family growth, holidays, vacations, rearranging furniture, and mood-swings.

Before you think you've got an original idea here, well, you do – with a nearly endless supply of suitable laptops being dumped on the market, a pile of fine woods and some stain … I think I just found something to keep me busy during retirement!

Chapter 8

Mobile Entertainment

It's about time we stretched our evil genius wings and made a reach for the open road. Just what exactly is there to do with a PC while rolling down the road in the middle of Utah or Nebraska or Oklahoma or Kansas?

No doubt you've seen many a 'mom van' and humongous SUV roaring down the lanes playing Disney or Sesame Street or Dora videos, wondering if the passenger kids are wide awake singing along or finally dozing off while Barney-tunes echo ominously in your head, wishing you'd gone for that upgrade option in your econo-box.

Now you can keep up with the Jones, Smith, Barnes, McTaggart, Anderson, and De La Croix clans by creating your own mobile DVD player system, one that will be much more portable, versatile and outlast the neighbor's three-year lease on their Euro buggy.

I tend to like projects that are either simple – functional, done and over with, or that have the potential for other things. This project shares both of those preferences in its simplicity, and its expandability.

You may already have the single largest piece of this project, or can use this as an excuse to obtain it – a laptop computer! Why use a laptop instead of going to the car audio shop or big-box discount place to have a built-in DVD player to match your interior?

First, you'll invest about $500 in a car DVD player installation – and there it sits – in the car. Second – with a built-in player you settle for a tiny 7- to 10-inch screen instead of a nice bright 12-, 14-, or even 15-inch screen. Third – yes, you can take it with you. A laptop can obviously be carted into the tent, camper, motel room, grandma's den or anywhere else entertainment is desired, and then it will not be a target for thieves. Fourth – it's a full laptop that can do many things besides displaying DVDs, especially on-the-road.

This is starting to get pretty obvious isn't it? Or are you second-guessing the twists and turns this can take? Oh, that would be cheating, but I am glad to see your thinking cap is warming up – especially with that last clue!

Simply having a laptop isn't enough, so let's move on to the parts list and a few options to proceed.

Parts List

- Laptop PC – must have a DVD-player and be able to run Windows XP Home or Professional with enough CPU speed, RAM and disk space to accommodate multimedia applications – 600–800 MHz minimum CPU speed (Pentium II or higher), 512–1024 MB RAM, 40 GB hard drive. Almost any laptop available in the last three years will fill this need, even if it needs a RAM upgrade.

- DC power adapter for above laptop, or DC power inverter for existing AC adapter. This could get a little tricky and is essential for this project. Most laptops sold for business – such as the IBM ThinkPad or Dell Latitude series – have a DC adapter, often called an 'auto/air' adapter available for them. There are some after-market models available but you need one with equal or better capacity ratings than the standard AC charger.

Optionally, you can buy a 100–140 watt inverter – a DC-to-AC converter/adapter from most electronics and even auto parts stores to run your laptop's existing AC adapter.

- DVD player software. Most laptops come with this, but if you get a used laptop from eBay, Craigs List or someone at a flea market or surplus store you may get no DVD software at all, or only an evaluation copy. Two good choices are InterVideo's WinDVD (www.intervideo.com) and PowerDVD (www.cyberlink.com) – one of which would come with the original laptop software packages.

- Headphones. No one can tolerate as much 'Barney' as a toddler and toddlers cannot read maps well enough to help you find the next turn.

- Headphone splitter. Sharing is a good thing, but not for headphones – every viewer should have a set.

- Cassette audio adapter or FM modulator to play DVD audio through your vehicle's audio system.

- Car mounting bracket/table. The web is full of these as finished products ready to install, from $30–100, but a trip to the hardware store and making your own is more fun!

Steps

1. RAM upgrade. If the laptop you want to use for this project has 256 or fewer megabytes of RAM you can upgrade most systems at anytime. Check to see how many RAM slots the laptop has, Figure 8-1. If the slots are already full – that will tell you if you need to remove two 64 or 128 MB modules and replace them with two 256 MB modules or one 512 MB module to get to 512 MB. Many older laptops, especially Pentium II and some Pentium III systems, do not support 1 GB of RAM (but this amount is highly recommended if the system will support it).

2. Windows. Installing Windows fresh, or using a current Windows installation is just fine. Even better if the DVD software is already installed.

3. DVD software. Installing software is a five-minute task consisting of clicking the Next button until the program's icon appears on your desktop and you may need to reboot.

4. DC adapter. Laptops are made to be portable and untethered, or typically static to sit on a desk being charged up, but mobile is a challenge between the two. Since most shows on DVDs are longer in duration than a battery in a laptop will run, you need to bridge the gap and feed your laptop's power-hungry needs. You could get a second or third battery, but they are expensive and when they are depleted your laptop is left cold and your passengers still wanting more entertainment.

Figure 8-1 *Installing and additional 256 MB of RAM in a laptop.*

Figure 8-2 *An example of a DC adapter for specific models of laptops.*

One choice is to acquire the DC or auto/air adapter made for your make and model of laptop, such as the unit shown in Figure 8-2. However, these units and most after-market all-purpose adapters are designed to charge the laptop battery with the laptop turned off, not as a permanent power supply, and they are useful only to power a laptop.

A more versatile choice is to get a DC-to-AC adapter or inverter, as shown in Figure 8-3, that converts the 12 volts of DC in your car to 115–120 volts of AC power that can be used for many things, including the AC adapter for your laptop. There are three advantages of this choice – one, you can use the AC charger you already have for your laptop; two, the laptop's AC adapter can power the laptop and charge the battery at the same time; and three, you can power other items from the adapter.

DC-to-AC adapters come in various styles, sizes, and ratings. You'll need to have a unit that can provide at least twice as much power as most laptop chargers are rated for – about 60–80 watts. Selecting a 120–150 watt unit will give you more capacity, and the unit will stay cooler and quieter (some units make a quiet to obvious squealing or buzzing sound when they are working.)

5. Headphone splitter. A laptop, and most computers for that matter, have only one audio output connector for speakers or headphones. The more passengers/viewers you have the more jacks you need. For two and not too much tangle a simple headphone 'Y' adapter, Figure 8.4, works fine.

6. FM modulator. Laptop audio circuits can provide adequate sound levels to two, and maybe even four sets of headphones, but the wiring becomes a real tangle. Also consider that your mini-van or SUV can seat how many? Six? Eight? Way too much wire – even for me!

Cyberguys.com sells a handy little modulator, powered from a cigarette lighter socket, that plays MP3 files from USB memory sticks, and includes a separate audio input connection on the side for feeding the audio from iPods or in this case, a laptop computer into the vehicle audio system through an unused FM stereo signal.

Figure 8-3 *A general purpose DC-to-AC adapter to run the laptop charger and other items.*

Figure 8-4 *A two-headphone splitter connected to the laptop headphone jack.*

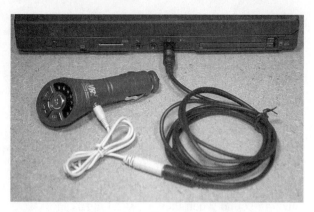

Figure 8-5 *Connecting the laptop audio output to an FM modulator.*

Figure 8-7 *Positioning the mounting base and adding on pipe sections.*

Figure 8-6 *Testing the laptop DVD player.*

7. Once you've chosen and assembled all of the gadgets, made sure the operating system works and the DVD software is installed, give the system a test spin, Figure 8-6, before you take on the task of mounting the laptop system and power cables in your vehicle.

8. Mounting system. There are many places on the web, from eBay to car stereo dealers, that will sell you a mounting system for a laptop, a DVD player, a PlayStation 2, an XM or Sirius radio, or various cell phones and PDAs. They all have merit but none I have seen allow you to remove some or all of the mount, without a wrench, to get it out of the way when you just want to ride in the car without electronics and hardware in your face.

To overcome some of the costs and limitations I created a hybrid mounting system using a six-inch tall cell phone bracket, a few easily obtained plumbing fittings, a scrap metal plate, and some Velcro™ strips to create a system I can disconnect and set-aside leaving almost no trace. By the way, as with most commercially available mounts, you will not have to drill any holes in your car.

The first part of this project is to size up the mounting location and positioning for the pieces. Use the base plate from the premade cell phone mount to mount the assembly to a seat mounting point on the vehicle floor, Figure 8-7, and work up from there.

As you build up the pieces of the mounting post you want to allow enough room to clear the top of the seat rail, and then the front of the seat when it is all the way forward. Once clear of the seat you

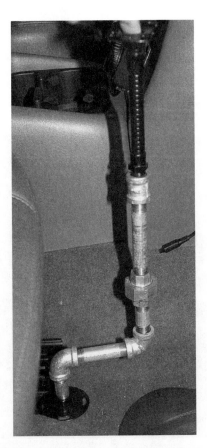

Figure 8-8 *The final pieces of the mount sized up and ready for top and paint.*

Figure 8-9 *The final pieces of the mount sized up and ready for top and paint.*

Figure 8-10 *The mount with laptop 'table' installed and painted.*

are free to build up to a comfortable level, including the flexible cell phone post and top bracket, Figure 8-8.

At the second vertical section past the seat rail I installed a pipe joint, Figure 8-9, which is the mechanism used to disconnect and remove the mount, leaving as little 'evidence' as possible. This joint could be closer to the mounting base but it would be harder to use so close to the seat rail. The joint also allows you to rotate and reposition the top after all of the pieces are tightened.

It is easy to underestimate the amount of vibration and loosening of fittings that happens in a moving vehicle – especially an SUV, mini-van, or truck. Before adding the top and painting the mount, tighten all of the threaded joints *very* securely using a strong pliers to grip and turn each piece. If the threads loosen up, the mount will become very unstable and cause excessive jarring of the PC and potential damage to the disk and DVD drives.

Next, prepare and attach a top plate made of sturdy material. I have a few pieces of scrap metal handy and found a very light weight but sturdy cast aluminum plate to use for my project.

Figure 8-11   *The mounting base installed.*

Figure 8-12   *The mounting post and table installed.*

Figure 8-13   *Velcro™ pads to hold the laptop on the mobile table.*

Placing it 'top' surface down, I located a center point then aligned the top mounting bracket, marked where to drill the mounting holes, drilled and counter-sunk the holes for flathead screws. You could use a piece of wood but that would be very bulky. Another alternative is to use a sheet of Lexan™ or Plexiglas™ available from some hobbyist, glass, or specialty stores.

With all of the pieces assembled, give the entire project a few light coats of gloss enamel or laquer (except perhaps a plastic top which does not accept paint very well). After allowing the paint to dry and set for 4–8 hours your mount is ready to install and use, Figure 8-10.

Installing the mount, beginning with the base section, Figure 8-11, should take all of 5–10 minutes depending on how easy it is to access the seat bolt and the tools available.

Figure 8-14 *Earthquake straps and brackets to secure computers and other equipment.*

The finished installation, Figure 8-12, will be ready for you to set your laptop in place. The PC installation may be accomplished by any of numerous means – typically the application of Velcro™ pads, Figure 8-12, or computer tie-down straps such as those used for earthquake bracing, Figure 8-13.

With the laptop secured to the mounting, Figure 8-14, you are ready to install the power cabling, optional car audio connection, and enjoy the show.

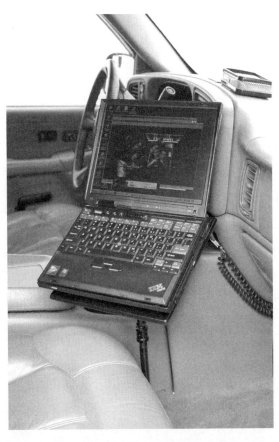

Figure 8-15 *The mount with PC installed playing DVD.*

Summary

As you've seen, this project is as much about preparing the installation as it is using a PC for mobile entertainment. If you can play DVDs you can certainly enjoy music as well by using the Windows Media Player or WinAmp programs to be the heart of a mobile jukebox. If you or your passengers are TV fanatics and miss not having access to soaps and talk shows while on vacation, you can add a USB-based TV tuner to this project and begin to have your own mobile entertainment center. However, entertainment and functionality come in many forms, as we will explore in the next project.

CAUTION! It is illegal in many states for the driver to have access to or view video displays, other than those used for navigation purposes, while driving. The mounting of this project demonstrates accessibility for a front-seat passenger but is similar and suitable for use in back seats as well. Drive carefully!

Chapter 9

Mobile Navigation

How often do you hear: 'Are we there yet?' echoing in the far reaching canyon of the station wagon, er, SUV. How many parents have suffered that inquiry without a clue about the distance or time to 'there'? (Remember when mom couldn't, or wasn't allowed to read a map, and dad would never stop to ask for directions?)

Here is a project that will keep dad out of the doghouse, allow mom to become a navigational wonder, and quiet that cavernous SUV by helping the kids learn geography and mapping.

Once the exclusive domain of hand-held GPS units, then in-car routing devices, mobile navigation systems have expanded into the typical PC with many affordable and fun products.

I've been a map enthusiast for a long time, by interest and necessity as I traveled much of the Midwest in various field service jobs. With a uniform interstate highway system and free maps at every gas station it didn't take long to learn the way to Marshall Minnesota or Grand Rapids Michigan. Directions and routes were one thing – familiarization with the terrain and local points of interest made us all modern day pioneers. As someone who is equally fascinated with mountain tops, or at least placing amateur radio systems on them, the terrain is very important – especially elevation and orientation in comparison to other geographic features.

There is still a 'wrong side of the tracks' which no map or database can tell you about, but finding the tracks is a lot easier these days. More impressive is the fact that today you can know precisely where you are in relation to the tracks, an historical attraction, the local hospital, and even the train station.

How is this possible? Well, it is probably a household term by now, but GPS, the Global Positioning System, is still fascinating and impressive for its functionality and the access we all have to GPS receivers and GPS-aware mapping software for our PCs. Every mapping software product on the market today, from DeLorme, Precision Mapping, and even Microsoft, is GPS-aware. Many include a GPS-receiver with the mapping software giving us easier access to explore this wonderful service.

You have a choice of software and GPS receivers to apply to this project. Microsoft Streets and Trips comes in many versions with and without GPS, as do DeLorme's Street Atlas and Undertow Software's Precision Mapping.

If you already own one of the popular Garmin or Magellan GPS receivers, Figure 9-1, you can buy a computer interface cable and re-use the GPS here with any of the available software. Indeed, your existing GPS can tell you where you are, but trying to plan a trip and actually see such a small display safely is a challenge.

Figure 9-1 *Hand-held GPS receiver with optional data cable integrates well with most navigation software.*

In the context of our PC projects, we have just introduced yet another excuse to have a laptop computer running in our vehicle – building our own GPS-based mapping and navigation system for far less than the usual GPS dash-top displays.

Parts List

- Laptop, DC power adapter, optional mounting system (from Chapter 8).

- Microsoft Streets and Trips 2006 with GPS Locator ($80–130), or DeLorme Street Atlas USA ($40–80) plus a GPS unit ($40–100).

- Serial-to-USB port adapter – if your GPS provides only a serial port connection.

- Optional interface to vehicle audio system to hear driving instructions from the navigation software – see the audio interface options in Chapter 8, Mobile DVD.

Steps

Once you have the basics in hand – a mobile laptop that is – this project is about as simple as it gets. For an incremental additional investment in a GPS unit and software your multimedia laptop just provided the foundation for what would have cost you $250–500 for a single-function navigation system that does nothing but, well, navigate.

1. The first step is to install the mapping software you prefer to use. For this project, though I have several software options to choose, I selected the all-inclusive Microsoft Streets and Trips with GPS Locator, Figure 8-2. As it turns out this is an excellent choice for ease of use and functionality.

Most products come in 2-CD sets, one containing the program and the other a library of maps, and offer the option to install the map data on your local hard drive so you do not have to keep the map CD in your drive – it is highly recommended that you do this.

2. Depending on your choice of mapping software, GPS unit, and available I/O port connection on your GPS and PC (serial or USB), you may need to install an optional adapter, Figure 9-3, and driver software to convert a serial port GPS connection to USB port on your laptop.

3. With the adapter installed (if needed), and GPS connected, configure your mapping software to connect to the GPS receiver. Some mapping programs will scan your system to find automatically the GPS. If yours does not know how to find a GPS by itself you will need to access Windows' Device Manager, Figure 9-4, to determine

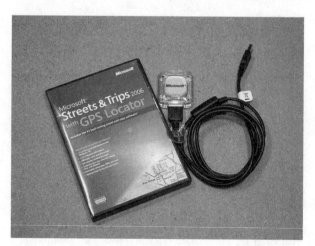

Figure 9-2 *Microsoft Streets and Trips with GPS Locator. The GPS receiver stays inside your vehicle.*

Figure 9-3   *An example of a serial-to-USB adapter required to connect some older GPS receivers to a PC.*

which COM ports are present in your system that may be used by the GPS.

The easiest way to access the Device Manager is to right-click on My Computer, select Manage, then select Device Manager. Expand the Ports (COM & LPT) listing to see the actual ports listing. When you know the available COM port

number you can try it in the mapping software GPS setup options, Figure 9-5.

4. With the software configured you are ready to try to acquire GPS signals and see where you are – literally – on your map. Your GPS receiver needs an unobstructed view to the sky above so it can pick up the signal from at least two, and preferably 4–6 available satellites orbiting the earth. Trees, rain gutters, roof lines, and even some forms of solar window tinting material can block or weaken the GPS signal.

Even a basic GPS receiver calculates its position internally, then sends the latitude, longitude, time, and even altitude data across the data cable to the computer. An example of this is shown in Figure 9-6, a screen shot from the GPSDIAG program. Much of the information is pretty cryptic or technical but it does make sense. Time, location, altitude, satellite information, and the raw data stream from the GPS receiver all correlate into useful data for your mapping program.

Once your receiver is tracking good satellite data you're on-the-road! Figure 9-7 shows the

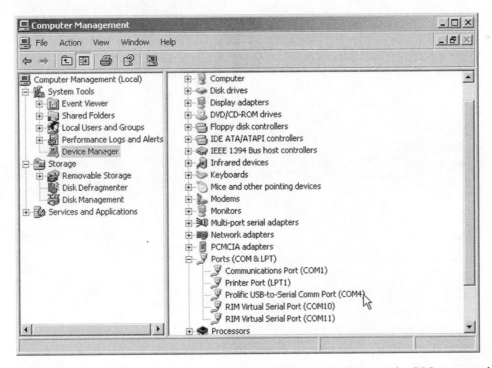

Figure 9-4   *'Windows' Device Manager showing available COM ports. In this case the GPS connected by USB cable assigned itself to be COM4.*

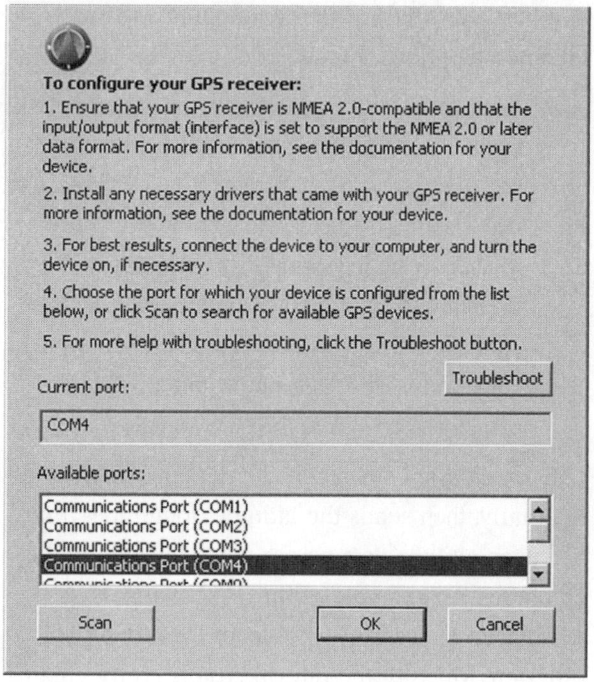

To configure your GPS receiver:

1. Ensure that your GPS receiver is NMEA 2.0-compatible and that the input/output format (interface) is set to support the NMEA 2.0 or later data format. For more information, see the documentation for your device.

2. Install any necessary drivers that came with your GPS receiver. For more information, see the documentation for your device.

3. For best results, connect the device to your computer, and turn the device on, if necessary.

4. Choose the port for which your device is configured from the list below, or click Scan to search for available GPS devices.

5. For more help with troubleshooting, click the Troubleshoot button.

Current port:

COM4

Available ports:

Communications Port (COM1)
Communications Port (COM2)
Communications Port (COM3)
Communications Port (COM4)
Communications Port (COM9)

Figure 9-5 *Microsoft Streets and Trips GPS settings dialog.*

entire PC and GPS installation tracking and displaying a current location. Once the mapping program is acquiring data you have several options to choose from.

First, you could just zoom into a map of anywhere the software covers – usually the United States, often Canada, sometimes South America or Europe – which is great because most of us can get a 'feel' for where we are or going to from a static map.

However, the purpose of any navigation system is to help you get from point A to point B, or let you create a route between two points to be used later. Streets and Trips facilitates these very easily. You have the option of obtaining route guidance – providing spoken and text indications of the directions you should follow to get someplace, or simply tracking from where you are to where you are headed – and capturing that information as a new route.

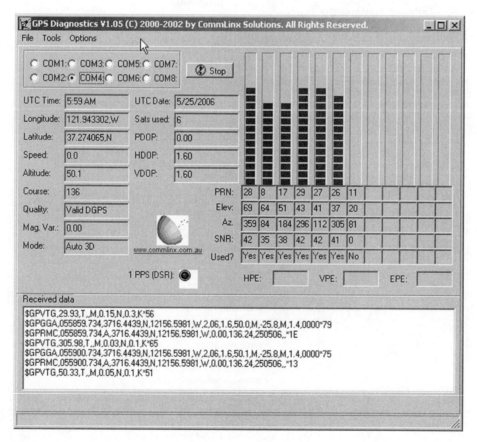

Figure 9-6 *GPSDIAG program displays raw data from your GPS receiver.*

Figure 9-7 *GPSDIAG program displays raw data from your GPS receiver, mounted inside the windshield.*

To start tracking a trip or obtain guidance, you must first tell the program that your current or other selected location is a starting point, as shown in Figure 9-8. Pick a point on the map, right-click, select Route, then Add as start… to begin your travels.

If you have an idea or specific address of a destination, type the address into the Routing edit box, then select Route Planner, Figure 9-9, to get the program to start telling you which turns to take and when. The directions will appear as text messages on the screen under the map and as voice prompts from the PC's sound system.

As directions come out of the PC sound system you will appreciate the sound-coupling recommendation in Chapter 8 – a laptop sound system is simply not loud enough to hear or understand voice prompts amidst the road and passenger noise in your vehicle. Playing the PC's sound through your vehicle audio system is much more effective.

Other features of note are the current location expressed in latitude and longitude in the upper left area of the GPS tracking screen, and the direction of travel which is provided not by a compass, but by calculating direction changes based on GPS data as the vehicle moves.

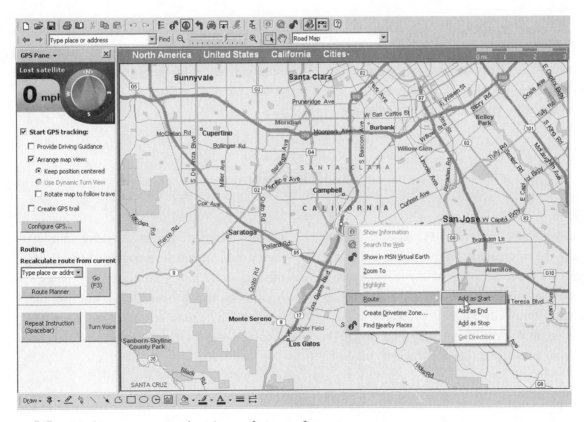

Figure 9-8 *Marking your current location as the start of a route.*

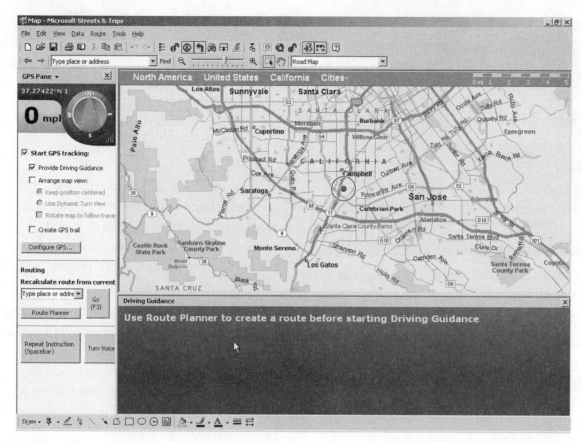

Figure 9-9 *Route planner gets you under way with directions to your destination.*

The PC-based map, and associated data, are vividly displayed on a 12-to-15 inch laptop display with much more clarity than any of the more expensive single-function GPS navigation products. As well, you can change the display mode to suit day or night driving, and highlight known points of interest (great for the kids who are simply dying to see the world's largest watermelon or a statue of Paul Bunyan).

Summary

You really cannot miss with a PC-based navigation system – the color, clarity, display size, and audio prompts are features worthy of any family vacation trip – be it simply from home to the local lake, or a full cross-country adventure. If you know where you are going you can always play a CD or DVD to lull your passengers (but not yourself) into peaceful distraction to let you drive your own merry way. Of course, a couple of things are missing – checking your e-mail and letting your friends and family 'watch' your trip over the Internet. There must be an answer to provide those capabilities as well.

Chapter 10

Mobile Internet

You're on the road, on vacation, away from it all, with that gnawing feeling you are missing something, forgot to tell the neighbor to water the begonias, or have to check on your eBay auctions (these feelings of anxiety typically go away after the third or fourth vacation day.)

Until you are calm, on your way to recovery from e-everything, your Internet-fix is just a plug-in away. Well, a plug-in wireless data card and a data service plan from your favorite cellular service vendor.

Cellular vendors offer many ways to stay connected – from Verizon's new Evolution Data Optimized (1X EV-DO) service to Cingular's General Packet Radio Service (GPRS) or High-Speed Downlink Packet Access (HSDPA), mobile users can experience nearly constant connectivity, even if only a meager 9600 bits-per-second, up to nearly low-end DSL speeds around 1 megabit-per-second.

Why cellular-carrier-based data plans instead of good-old public WiFi networks? Simple – cellular carriers are ubiquitous – even in rural areas the cellular carriers provide voice and at least modest data coverage. WiFi is still largely unknown or useless beyond a few square blocks of urban population, and disappointing at best in rural areas where there may be no DSL or cable Internet services.

As well, cellular carrier services use a single connection program and sign-on process no matter where you travel. WiFi services may be totally open and free, open but encrypted if you know the local provider, or exclusive to any one of several WiFi subscription services. To stay covered by WiFi you would have to stay within urban or suburban areas and pay subscriptions to multiple WiFi services.

Another very significant difference is that WiFi is not designed to maintain a connection when traveling at high rates of speed or for significant distances – in the time it takes to travel the distance of a couple of city blocks your connection is gone never to be seen again. With cellular data services, like voice, connections are designed to survive travel between cellular radio sites. (Yes, we can hear you now!)

Even if you are trying to escape technology while you are on vacation (wait – why the heck are you even reading this book – get packing!) there are some pretty good reasons to maintain a high-tech coupling to the rest of the world – especially while traveling, not just when you are checked-in to your hotel.

As you travel down the road with your mobile navigation system, living the carefree, no plans, no expectations vagabond lifestyle you may still need to find a place to eat, sleep, get gas or recharge your fuel-cell. No mere map, atlas or motor club travel guide can tell you precisely where the next services are ahead, or provide the ability to reserve a room while zooming down the interstate or bouncing along an unpaved country road. Do not forget being able to check the closing time for the world's largest watermelon museum before you detour off your main course or zoom right past it – next exit 30 miles.

With truly mobile Internet service you could also leverage your Vonage or Skype Internet-phone services (if you don't have cell phone voice service), of course send pictures of your trip to friends and relatives, even include a webcam – and yes – check your auctions and stocks, and eventually send GPS information about your location to the web so folks could 'watch' you travel cross-country.

Parts List

- Mobile-capable laptop.

- Verizon or AT&T/Cingular mobile data card and service agreement.

Steps

1. Install the connectivity and driver software for your mobile data card and service. Reboot your laptop as indicated by the software.

2. Install the data card, Figures 10-1 and 10-2, allow it to be recognized and the driver software properly loaded for it.

Figure 10-2   *Cingular's second generation GPRS/HSDPA wireless card.*

Figure 10-1   *Verizon's IX EV-DO wireless PC data card.*

whatever you need or want to do over the Internet while traveling.

3. Test the wireless data connection, Figure 10-3 and access to your usual Internet sites. If you can access Yahoo!, eBay, MSN, Google, etc. then you're all set to go for

Figure 10-3   *Active Internet connection through Cingular's GRPS network.*

4. Depending on how much you need to rely on your new mobile connectivity, you may want to acquire an external antenna for your data card to provide better range and more reliable connectivity as you move about. Browse to www.antennasplus.com to find an antenna that suits your needs.

Summary

So far, adding a modem and a piece of software to your laptop is not a very skill- or time-challenging project, but providing high-speed connectivity opens up many possibilities to exploit your mobile and at-destination on-line interests when away from the comfort of cable or DSL.

Some of the obvious, and not so obvious possibilities that have been mentioned and may be considered are:

- Voice-over-IP (VoIP) applications.
- Instant messaging.
- E-mail.
- Webcam.
- Looking up attractions relative to your travels, aided by GPS mapping software.
- Weather tracking via SwiftWx (www.swiftwx.com).
- Sharing your location through GPS tracking.
- Uploading your digital photos without stopping to find a place for WiFi or dial-up.
- Providing shared-Internet access through WiFi connections in isolated areas.

I'll share projects related to a couple of these ideas in the next chapter.

Don't want to buy a new card? Check with your cell phone provider to see if your phone provides Internet data services. In either case – card or phone – beware that data connections can increase your cell phone bill significantly unless you have a preset data plan added to your service.

Chapter 11

Share Your Trip With APRS

As you're cruising down the road, with your passengers watching DVDs, navigating with your GPS software, and wandering the Internet looking for places to eat or sleep, you may have thought about how you could share your travel progress with others.

For about $350 to buy a GPS-tracking system, plus a monthly subscription fee for tracking services, you can have your whereabouts reported and tracked by one of many vehicle and fleet tracking services. This cost is quite steep if you already have most of the components and smart bits already, and you just want to occasionally broadcast your trips.

If you've followed and built the previous projects, you already have most of the 'stuff' to make your own tracking system, minus a radio transceiver to get your trip data from your car to the Internet. This project also requires you join thousands of amateur radio ('ham') operators who share their travels through the Amateur Position Reporting System (APRS) technology, by getting your 'ham' license.

APRS technology allows its users to send GPS location data over the airwaves to any of hundreds of receiving stations that relay the data to the Internet and many APRS-tracking sites like www.findu.com. APRS is not merely a geek's novelty – it provides valuable data in times of emergency – providing a significant (free) service to assist wild-land fire crews, search and rescue teams, and the National Weather Service with severe storm tracking.

It is also possible for non-'hams' to send the same GPS position information over the Internet in the form of weather information reports. Weather? Yes, there is a voluntary project called the Citizen Weather Observer Program (CWOP) that allows location data and weather station information to be reported over the Internet to and through APRS servers, and viewed with a web browser.

Implementing either method of sharing GPS data involves many of the same steps, though, if you have any interest in technical pursuits, I encourage you to look into amateur radio as the ultimate experimenter's realm. Most of the great technical innovations of 'our time' can be traced to their origin or significant participation by engineers, scientists and technicians who have extended their experiments and concepts through amateur radio – much like the open-source software and computer hacking hobbyists more easily recognized today. Combining amateur radio and computers makes for a tremendous range of experiments, worthwhile projects, services, and career opportunities.

Two of the most interesting APRS tracks I've observed were from a fellow ham and friend broadcasting his position along an Alaskan cruise, and his cruise through the North and Baltic Seas from London to St Petersburg Russia. Such ventures and technology have changed the notion of postcards and telegrams into real-time 'we are here!' vicarious adventures.

First, I'll cover the the GPS-via-APRS/radio method, software, hardware, radio-to-computer wiring, and all, then, in the next chapter, sending GPS location data as weather data over the Internet, from a mobile device. Two very exciting projects to put yourself 'out there' quite differently than blogging or a webcam.

There are many ways APRS users get GPS or computer data back and forth over radio – the most common is to use a 'packet modem' – a device

Figure 11-1 *The Kantronics KPC-3 is a very popular data-over radio packet modem.*

Figure 11-2 *The TinyTracker 3 packet modem is used to connect a GPS to a radio.*

much like the internal or external data modem you would use for dial-up Internet access.

This project involves installing and configuring multiple pieces of software to integrate your PC's sound card capabilities as a software-based packet modem, provide software-to-software connectivity between the sound card's new packet modem function and the data program that reads and sends GPS location data. When all the software is working correctly, you will connect your PC's sound card and serial/COM port to your radio, make a few adjustments, and then be able to 'see' yourself on Internet mapping sites.

A typical packet modem, Figure 11-1, costs between $120 and $250. Like a data modem it accepts information and commands from a PC's serial/COM port, then converts that information into audio signals for transmission over voice 'circuits' (a radio transmitter), and a control signal to turn a transmitter on or off. As we go through this project you'll also see that you can continue to use your GPS for your local mapping and navigation while sending your position via radio.

Another way is to use a packet modem, Figure 11-2, specifically designed to be a GPS-to-radio interface to be used only for tracking over radio or the Internet.

A few very clever programmers figured out that the sound card in a PC, and the serial/COM or parallel/LPT port could do the same thing without the need for a separate packet modem. Since we already use a PC to read and type information we are exchanging, being able to send and receive over radio just became more economical.

The software that makes a PC and its sound card into a packet modem allows many types of programs to interact with data received from and sent over a radio – from text messages, e-mail, weather information, and for this project, location data from our GPS.

It takes about an hour to download and install all of the software necessary to do this project, perhaps an hour to get to and from your local electronics store for some parts, another hour to put the parts together, a half hour or so of testing, and then you are ready to roll, sharing your location as you go.

About the Software

The five pieces of software used to make this project work have been created by hobbyists who saw both opportunities and challenges to advance technology – typical of amateur radio operators whether they are building equipment, experimenting with antennas, or sitting at their keyboards considering ways to integrate communications tools and radios.

The first piece of software – 'AGW Packet Engine' or 'AGWPE', was written and is maintained by George Rossopoulos, amateur radio operator callsign SV2AGW, in Greece. AGWPE is but one of many pieces of software George created to facilitate many uses of packet radio. This piece of software provides three very important features for our project:

- Turns a PC sound card into a packet modem to send and receive the warbling tones sent via radio, converting them into useful data – and for not just one but two possible radios!

- Provides data input and output from the 'sound card packet modem' functions via TCP/IP, within the local PC and available over network for other PCs.

- Provides control over the data signals on a serial/COM or parallel/LPT port to turn on a radio transmitter.

The significance of creating a TCP/IP-accessible radio modem is tremendous – allowing multiple programs and even multiple computers to use the radios on a shared basis. AGWPE replaces one or more radio modems by leveraging the capabilities of PC technologies. To go with AGWPE you need the TCP/IP drivers for your operating system – from the AGWTCPIP.ZIP file on George's site.

The second piece of software we need, KipSSE, bridges the TCP/IP interface of the AGW Packet Engine to the TCP/IP interface of our data communications and GPS processing program, APRSplus. KipSSPE makes AGWPE look like a hardware modem to APRSplus.

APRSPlus, or APRS+SA as it is seen on screen, was written by Brent Hilldebrand, amateur radio operator KH2Z. Its function in our project is to collect GPS data, combine it with data specific to the APRS network and make it presentable to a radio or network interface to be received by other amateur radio stations or APRS servers. APRSPlus will also accept and present APRS location and other message data over radio or the Internet, completing the ability to have two-way message communications through multiple mediums. APRSPlus communicates with KipSSPE, sending the GPS+APRS information that eventually ends up transmitted over radio. APRSPlus can also read weather station data files and forward weather information around the globe.

APRSPlus also has a logical interface to some older versions of the DeLorme Street Atlas mapping program so it can actually plot and present the locations of other APRS and weather stations. Without KipSSPE, AGWPE, and the sound card and COM port interfacing to a radio, APRSPlus would have to be connected to a packet modem through a serial/COM port (or USB-to-serial adapter).

The last piece of software we can bring into this project is GPSGate, the product of Johan Franson, founder of Franson Technology AB. Johan is not an amateur radio operator but his software represents the spirit of making things practical and functional with PC technology. GPSGate is software that creates a logical serial/COM-port splitter within your PC allowing two or more programs to use the data from one GPS – in this case your mapping software and APRSPlus.

Since much of this software is Shareware and has a limited evaluation period, please be sure to register the products. After the software is installed and configured, we have a little hardware work to do – connecting the PC to a radio transceiver, and we're on-the-air!

Parts List

- Amateur radio license – this will cost you a weekend or a few evenings at a training class or self study then taking a test, about $15–35, and will return the pride of accomplishing a very significant step towards a new world of technology and service. More info and sample tests on-line at www.arrl.org

- PC with sound input and output, and serial (COM port) I/O capabilities (yes, a USB-to-serial converter will work).

- GPS unit with serial or USB I/O connection – Altina, Microsoft, DeLorme, Garmin, Magellan, etc.

- Amateur radio VHF transceiver – almost any 'rig' from an old Icom IC-2A to a current Kenwood TH-D7 hand-held radio will work fine. Cost is between $40 and $400.

  - You'll need a car charger/power cord or extra battery packs, and an external antenna for your car is recommended.

- Connectors and wire to build an interface cable between radio and PC:

  - DB-9 female (serial port) connector

- 1 each 3/32 inch and 1/8 inch stereo plugs – for PC

- 1 each 3/32 inch and 1/8 inch mono plugs – for radio

- A 'couple of resistors and capacitors and transistors' from Radio Shack.

- AGWPE software from http://www.elcom.gr/sv2agw/agwpe.zip

- AGWPE TCP/IPdrivers from http://www.elcom.gr/sv2agw/tcpip.zip

- KipSSPE from ftp://ftp.tapr.org/aprssig/winstuff/aprsplus/KipSSPE.zip

- APRS Plus http://www.tapr.org/~kh2z/aprsplus/

- Franson GPSGate software from http://www.franson.com/gpsgate/

Tools

- 'Typical' hand tools – needle-nosed pliers, wire cutters/strippers.

- Soldering iron and electronic (60/40- or 63/37-blend rosin core) solder.

- You can avoid the tools, connectors, and soldering if you buy a premade radio-to-PC interface unit and radio-to-interface cable from www.buxcomm.com

Steps

1. Install the AGW TCP/IP driver appropriate for your operating system. This is done by opening the Windows Control Panel, then selecting Add Hardware. Starting the Add Hardware wizard leaves you prompted for the state of the new hardware, as shown in Figure 11-3. Select 'Yes, I have already connected the hardware' then click the Next button.

The wizard will present a list of devices it knows are installed, Figure 11-4. Scroll to the bottom of the list, highlight 'Add a new hardware device' then click the Next button.

The wizard will prompt for a selection of automatic or manual new device selection, Figure 11-5. Select the "Install the hardware that I manually select from a list (Advanced)" radio button then click the Next button.

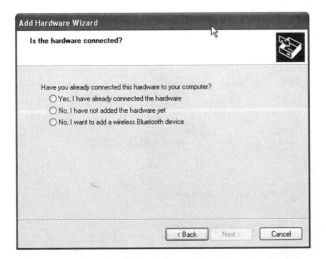

Figure 11-3   *Starting Windows' Add Hardware wizard to install the AGW TCP/IP driver.*

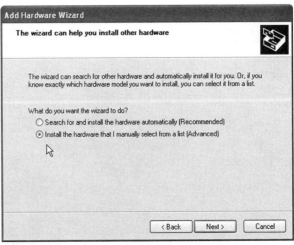

Figure 11-5   *Selecting manual installation to get to the TCP/IC driver.*

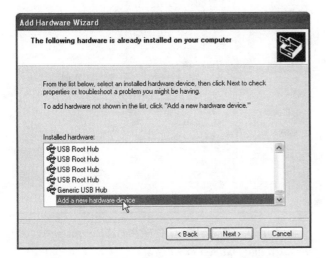

Figure 11-4   *Choosing to add a completely new hardware device in Windows.*

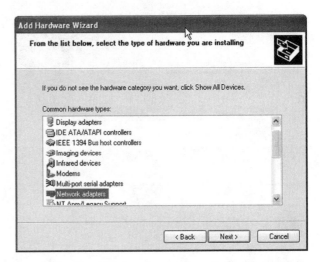

Figure 11-6   *Selecting network device shortens the time it takes to find a driver from a disk location.*

To save a little time, scroll down the list to and highlight Network adapters, Figure 11-6, then click the Next button.

When the Select Network Adapter lists appear, Figure 11-7, simply click the Have Disk… button, then browse to the location of the AGWTCPXP.INF file (from the AGWTCPIP.ZIP file you downloaded earlier), Figure 11-8.

Click the Open button, then the OK button until you see the SV2AGW Drivers in the list, under Manufacturer, Figure 11-9. Highlight SV2AGW on the left, then highlight SV2AGW TCP/IP Over Radio NDIS Driver on the right, then click the Next button.

Follow the remaining prompts to complete the installation of the driver. If Windows pops up a warning that the driver has not been signed/certified, simply click the Continue Anyway button. With success, you should be able to right-click the Network Places icon on the Windows desktop, select properties, and see the SV2AGW TCP/IP entry, Figure 11-10.

In most cases, and until you're ready to take on some advanced steps such as sharing your radio interface across your home network, or the Internet, you are done with the AGWPE TCP/IP driver and can proceed to install the Packet Engine.

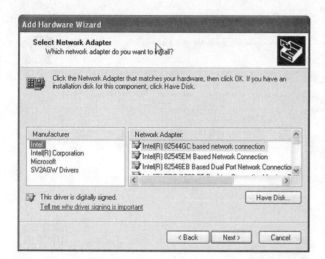

Figure 11-7 *Click the Have Disk… button to navigate to the location of the driver file you want.*

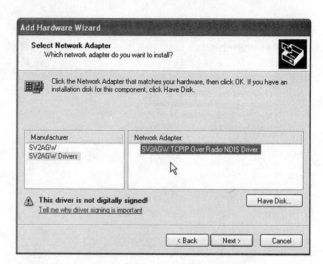

Figure 11-9 *Selecting SV2AGW Drivers and the SV2AGW TCPIP Over Radio NDIS Driver.*

Figure 11-8 *Navigate to the location of and select the AGWTCPIP.INF file.*

Figure 11-10 *The SV2AGW TCPIP driver properly installed and enabled in windows.*

In the RadioPort Selection dialog, Figure 11-12, select the New Port button. This will bring you to the settings dialog to create your new software and sound card-based 'radio port', Figure 11-13.

The first settings area, Select Port, is where you specify the serial/COM port on your PC that you will use to control your transmitter. For laptops with a single serial/COM port this will probably be COM1. For laptops with both a serial/COM port and an internal modem, this may also be COM1, but could be COM2. If your laptop has no

Figure 11-11 *The AGWPE icons appear in the task tray when the program is running.*

2. Unzip the AGW Packet Engine software to a known folder on your system – I typically use C:\Program Files\AGWPE – then create a shortcut to the program on my Windows desktop. Double-click the program shortcut. You will see a splash screen and then notice only a new icon in the Windows task tray, Figure 11-11. Click the new icon then select Properties to begin the setup of the Packet Engine.

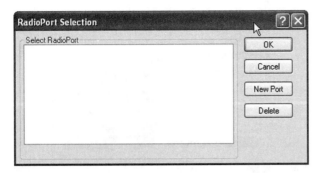

Figure 11-12   *AGWPE before port configuration.*

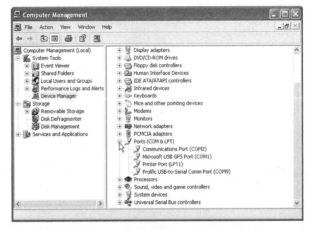

Figure 11-14   *Viewing available COM ports in Windows' Device Manager.*

Figure 11-13   *The COM port settings for AGWPE's sound card-based packet modem.*

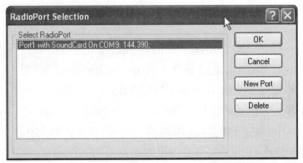

Figure 11-15   *AGWPE showing a successful port configuration.*

serial/COM port connections and you find yourself using a USB-to-serial port adapter.

To see which ports exist on your system, right-click My Computer, select Manage, then select Device Manager in the left pane, scroll in the right pane then expand Ports (COM & LPT), which will reveal details similar to those shown in Figure 11-14. In this example we see a Communications Port (COM2) which is most likely a built-in serial port, or perhaps the internal modem. Knowing this particular system, it does not have a serial port connector so this must be the internal modem. Instead, I've used a USB-to-serial port adapter, which appears as Prolific USB-to-Serial Comm Port (COM9). This is the value to put into the Select Port area above.

The default speed of 9600 baud is fine, since the interface to the radio uses only one of the signal lines from the COM port and there will be no actual serial data transfer on this port. There are three parameters to configure:

- Under TNC Type, select SoundCard.

- To the right, under Tnc Control Commands, for now, select the SinglePort radio button.

- At the bottom of the dialog, under Tnc RadioPort, for Port1, simply enter the frequency to be used for the APRS data transmissions – 144.390, and leave the rest alone. This value does not set the frequency of the radio, it simply gives me a reference to what is connected to the port.

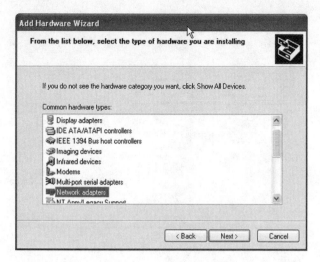

Figure 11-16 *Enabling TCP/IP over radio, though it is not required for APRS.*

Figure 11-17 *A typical configuration for the KipSS/PE middleware program.*

Click the OK button to complete this setting, then, as indicated by the program, close AGWPE – do this by clicking the AGWPE icon in the task tray then select Exit. When the icon disappears, restart the program from the desktop shortcut to adjust a couple more settings.

When the icon reappears, click on it then select Properties to verify your new RadioPort has been established, as shown in Figure 11-15.

Click the AGWPE icon again then select TCPIP Over Radio Setup to access the Dialup<>AX25 Setup dialog, Figure 11-16.

Within this dialog there are four items to configure:

- Click the Enable TCPIP radio button.
- Enter the registration code for AGWPE (you did register the program?).
- Select the Port1 previously configured above.
- Then enter your amateur radio callsign.

Click the OK button to save the data and close the dialog. Shutdown then restart the AGW Packet Engine one more time. The TCPIP driver and the Packet Engine are now ready to act like a radio modem, accessible from other programs. There are two optional Startup Programs items we can

configure later to make it easier to run everything with one click.

3. Unzip then copy the KipSSPE program and two supporting files to your C:\Program Files\AGWPE folder. Double-click the KIPSSPE.EXE program to start and access its single configuration dialog, Figure 11-17.

In the KipSS/PE dialog, only five items need configuration:

- Settings should reflect the localhost IP address of 127.0.0.1 with a port of 8000 (expressed as 127.0.0.1:8000).
- Click the Open button so it says PE Close.
- Lower in the dialog the IP address should also be 127.0.0.1.
- Check the Auto-Start box.
- Click the Open TCP/IP button so it says Close TCP/IP. Click the 'X' in the upper-right corner to minimize the program to the task tray.

4. Install the latest version, 2.28, of the APRS+SA program, which has a more conventional Windows installation dialog to follow. Once installed, run the program, note the unregistered version splash dialog, then proceed to setup the program parameters by selecting Setup then Main from the main screen, Figure 11-18.

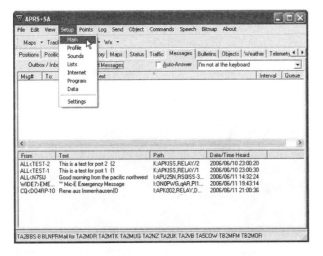

Figure 11-18 *Beginning the setup of APRS+SA.*

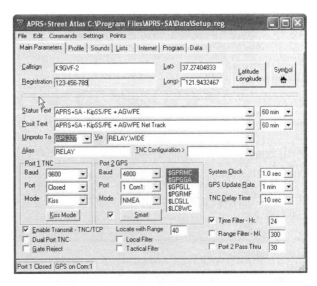

Figure 11-19 *Typical settings for the main parameters of the APRS+SA program.*

For Main Parameters, Figure 11-19, enter or set the following:

- Your amateur radio callsign, followed by a station subidentifier to indicate you are a packet radio station – using -1 is common, but -6 or -9 designate mobile operations.

- Your longitude and latitude, which can be manually entered or captured from your GPS if you select that Latitude Longitude button.

- Your registration code for the software.

- Ignore Port 1 TNC.

- Set Port 2 GPS to, typically 4800 baud for most GPS data streams, the COM port your GPS is plugged-in to (without your GPS mapping program running, more on this and using Franson GPSGate later).

Select the Sounds tab, then deselect ENABLE WAV SOUNDS, as the voice and various sounds from this program will interfere with the data modem functions of the sound card.

Select the Internet tab, then enter the following parameters as shown in Figure 11-20, and select the following parameters:

- Click the UDP-Open button to make it change to UDP-Close.

- Click the Open-KipSSPE on startup checkbox.

- Select or enter the IP address of 127.0.0.1 to the right of the Open-KipSSPE button, then click the button to make it change to Close-KipSSPE.

Click File on the menu bar then select Save to save the parameters to the setup.reg file. (Saving the settings is not possible with an unregistered version of the program.)

This completes the basic setup of the APRS+SA program. We're now ready to test the GPS functions of the program, Internet connectivity to get a taste of what the program can do, then move

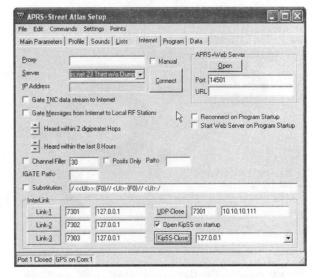

Figure 11-20 *The Internet settings for APRS+SA.*

Coordinate Conversion and Map Capture

Hide	Deg°	Min'	Sec"	Capture
Latitude	37.27396			Reset
Longitude	-121.94315			OK
	Zone	Easting	Northing	> Map
LL > UTM	10S	593697	4125788	UTM > LL
Datum	WGS84			Datum
LL > GS	CM97AG65TS70GC73			GS > LL
LL>APRS	13716.44N/12156.59W/			
Decimal Degrees°				

Figure 11-21 *APRS+SA will establish its first location automatically from your GPS.*

on to connecting a radio to the sound and COM port connections.

5. If you have not already discovered that APRS+SA will read your GPS data through the Setup/Main/Main Parameters/ Latitude-Longitude dialog, Figure 11-21, that's the first screen you can use to check things out. If your GPS is reporting to the program properly your location will be filled in for you.

If you want to see the stream of raw data from your GPS, go to the GPS tab on the main program screen. Once per second you should see a new set of data scroll through, as Figure 11-22 shows, for example.

6. Interface your PC to your radio. To make things a bit simpler, I purchased a single

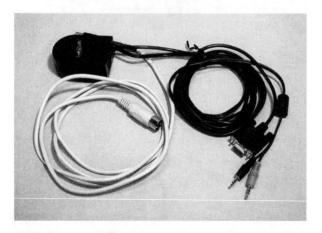

Figure 11-22 *The Bux Comm RASCAL PC-to-Radio Interface module.*

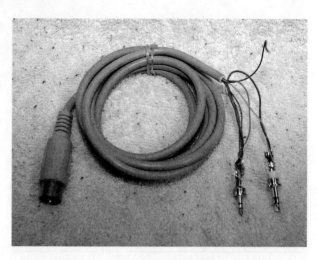

Figure 11-23 *A hand-made interface cable for the Bux Comm RASCAL and an Icom portable radio.*

RASCAL radio-to-sound card interface unit, Figure 11-22, from Bux Comm Corporation (www.buxcomm.com) with one set of pre-made cables for my 'HF' radio to save time for another project.

Since the interface unit is somewhat universal and can be fitted with a new cable built for any radio, I simply scrounged an old 5-pin DIN plug keyboard extension cable, clipped off the female end I didn't need, and added my own connectors to plug into one of my Icom IC-2AT VHF radios for use on the 144.390 MHz national APRS frequency.

The resulting cable and connectors for the Icom radio, left unfinished (without cleaning up the wire dressing and connector bodies) for display here, is shown in Figure 11-23.

The schematic for this cable is shown in Figure 11.24. A schematic for a completely home-built interface you can make with parts from Radio Shack is shown in Figure 11-25.

7. Once the interface cabling is built you need to test it to be sure it does three things:

   • Allows the sound card packet radio program to receive and decode signals from other transmissions.

   • Keys the radio transmitter so you can send APRS position data to others.

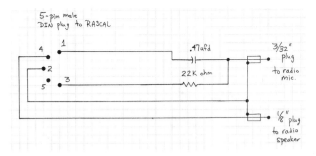

Figure 11-24 *Schematic of the hand-made Rascal-to-Icom radio cable..*

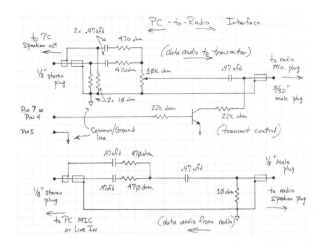

Figure 11-25 *Schematic of a generic PC-to-radio interface cable.*

- Sends radio modem tones through the transmitter so other stations can receive your APRS data clearly.

Simply connect the cables to the right places and start testing for functionality. The tricky thing about the BuxComm RASCAL interface and the AGWPE software is that you have no control over which audio channel is used by which radio port in the software, so you may have to experiment and swap port assignments. In one case I had radio Port #1 transmitting but without audio, and radio Port #2 getting audio but not transmitting. I ended up reversing the audio cable wires inside the BuxComm interface unit and all is well with Port 1 both keying and sending data. The hand-built interface from Figure 11-24. above is intended for one radio port use only, but could be modified to split Left and Right audio channels for two radios.

For testing the first item, the AGWPE program provides a sound card adjustment tool that will show you the waveforms and patterns of incoming signals to help you adjust the radio's volume control and your sound card Line In or Microphone level for a nice 'pretty' waveform. Simply click the AGWPE icon in the task tray and select Sound Card Tuning Aid. Instructions with the program as well as the help file are very good references to help you understand what you are looking at. (This little feature alone is fun to play with as your PC becomes a multi-function oscilloscope to analyze radio signals.) Your radio should be able to receive at least a few other nearby APRS stations (unless you live in the

middle of a wilderness area), but otherwise the program will display the pattern of random white noise by which you can adjust the level controls so there are not flat/squashed waveforms visible.

For the second, the APRS+SA program can and will be used to control your radio transmitter. Observe the transmit indicator on your radio, or listen with another radio, in the APRS+SA program, then press the Ctrl and P keys at the same time – this will cause APRS+SA to key the transmitter and try to send out a data stream. You should see the transmit indicator on your radio light up. If you are listening on another radio tuned to the same frequency it should indicate at least a received signal is present. Or you may also hear the burst of tone signals that make up the APRS location data stream.

If in the second test you hear tones on the other radio but they are very low in volume, this is an indication that the sound card speaker/headphone volume or transmit level adjustment on the PC-to-radio interface needs to be adjusted to produce higher level audio output. If the tones you hear are very loud and distorted you need to reduce the volume level until the tones sound, well, if not pleasing, at least 'clean'.

This is truly not a scientific test of or means to set the transmitter modulation level. For this you need to use a deviation meter, deviation scope, or

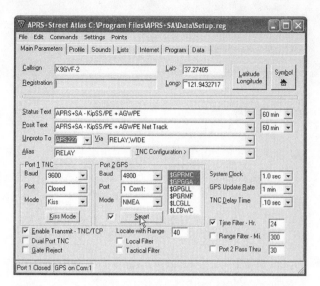

Figure 11-26 *Selecting the SmartBeacon settings in APRS+SA.*

communications service monitor – which a nearby 'ham' may be able to help you with. Measuring the modulation level with test equipment is the best way to make this adjustment. The second best way is to work with another station within range to determine if it can receive and properly decode the APRS location data you are transmitting. If it can, your level is set pretty close, if it cannot the level is too high or too low.

Once you are sending and receiving APRS data the only steps left in the tracking project itself are to make one last adjustment in APRS+SA so your GPS data is sent 'just right' for your travel speed and adequate tracking without flooding the radio systems and Internet servers with your data reports, and to put the whole system in the car and take it for a spin.

8. Configure and enable the 'smart beacon' settings in the APRS+SA program so the system sends your position automatically at regular intervals, allowing people to track your location and movements. From the main menu bar select Setup then Main. In the main setup dialog, under Port 2 GPS, click the check box next to the Smart button, Figure 11-26.

Click the Smart button to access its parameters, Figure 11-27. There are seven adjustable parameters:

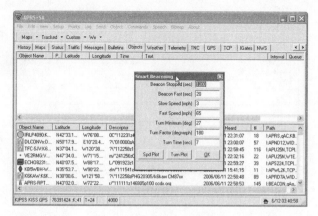

Figure 11-27 *SmartBeacon parameter list for APRS+SA.*

- Beacon Stopped (sec) – the lowest value is 1200 seconds = 20 minutes. When the program calculates that you are not moving it will transmit your position at this interval.

- Beacon Fast (sec) – the interval at which the program will send position data when you are moving faster than the Slow Speed (mph) setting, but slower than the Fast Speed (mph) setting.

- Slow Speed (mph) – the speed above which movement is deemed valid, below which the program determines you are stopped and sends a beacon at the interval set by Beacon Stopped, above.

- Fast Speed (mph) – the speed at which beaconing interval times no longer change with speed so some tracking effects will be less accurate. In some APRS programs beacons stop above this speed, which I suppose reduces the evidence of speeding.

- Turn Minimum (deg), Turn Factor (degxmph), Turn Time (sec) – these three parameters control how accurately a turn is tracked versus beacon timing and the appearance that a tight turn may look like a stop. These are best left at default values.

- The Spd Plot and Turn Plot buttons will display graphs of collected data points.

SmartBeaconing[tm] was invented by Tony Arnerich KD7TA and Steve Bragg KA9MVA to balance the number of beacons sent versus speed and turns to provide a more accurate tracking of a vehicle in motion than simply sending out positions based on time.

Sharing Your GPS Data Between Mapping Program and APRS

One problem you will encounter with PC serial ports, and programs that try to use them, is that they do not share well, if at all. Once you hook up your GPS to a serial/COM port and run a program to map data from it, only that one program can accept data from or send data to that port – which is fine if you have only one program that needs the data.

As we build up our navigation system we find that we probably want two programs to acquire data from our GPS receiver – our mapping program and our APRS program. There are a couple of ways to work around the one-port/one-program problem – one is to use or add a second serial port and wire up some widget to split the data from the GPS to go to both ports. This is cumbersome because most laptops do not have two serial/COM ports, or you end up with even more wires to get tangled in.

The second option is purely logical – let one program deal with the serial/COM port and GPS, then split the data it receives between two or more other programs. This is possible, at least in Windows, because physical serial port signals are made available to programs only as device driver programs anyway – so if we insert another program that acts like a serial port device driver, or many of the same, we can create multiple logical or virtual serial ports out of one. In fact that is what the AGWPE program is doing – making a software driver for the sound card act as if it were a serial port and hardware packet modem.

The answer is a quite novel piece of software, GPSGate – written specifically to share GPS data

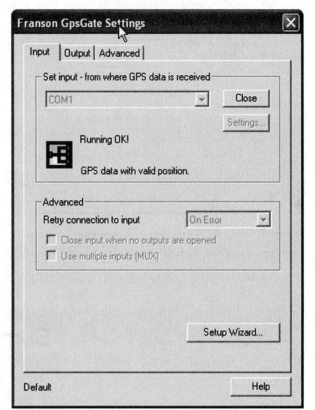

Figure 11-28 *Typical input port configuration of the Franson GPSGate program.*

across multiple programs, it could be used for other things, but it's perfect for our project. GPSGate's function 'own' a specific serial/COM port, as the one and only program that can do so, then it can be configured to spawn multiple virtual ports to be used by multiple programs – perfect! This imposes a couple of changes to our mapping and APRS software, but nothing dramatic.

First, download (www.franson.com/gpsgate) then install GPSGate. There are two versions – GPSGate Express is designed to share one GPS

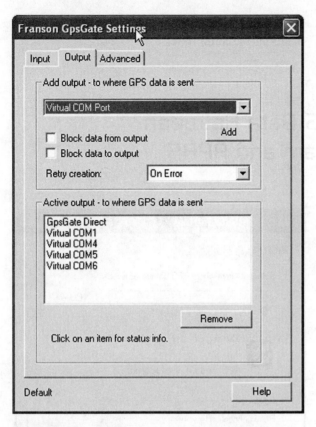

Figure 11-29 *GPSGate output port options showing four virtual and one network-based COM Port available.*

with only two programs, GPSGate Standard can share one GPS with any number of virtual COM ports and over TCP/IP connections.

The setup wizard with GPSGate Standard will scan your system for recognizable GPS data on all available COM ports and attach itself to that port, Figure 11-28, then it creates both a TCP/IP instance and four Virtual COM ports, Figure 11-29, to share data with from the incoming GPS stream.

You may of course reconfigure the program for any incoming COM port and any number of outgoing ports; as well, you can run the program many times for multiple GPS input and more output ports.

Once you have the output ports determined or reconfigured, those ports are available to be used by your other software programs. Simply change the COM port selection in your mapping program to one of the available GPSGate output ports, and your APRS program to use another GPSGate port and you're all set!

Summary

There is a certain mystique about transmitting into the 'ether' and realizing the results coming back in the form of a blip on a Google map over the Internet, knowing someone in a far away place may be tracking our trip across the Plains or down

a scenic coastline. The Internet and to some extent amateur radio allows us to establish at least vicarious if not long-term relationships with the people and world around us.

Chapter 12

Let Others Track Your Trip By Google

Working with a GPS and mapping program is a lot of fun – it shows us a lot about geography, new places and things to discover, and keeps us from getting lost. Being able to share your trips with others has previously been limited to amateur radio operators using APRS technologies, or customers of commercial (expensive) vehicle tracking services – until now.

Using a GPS receiver, a laptop (we'll call it a 'car-PC'), a cellular data modem or WiFi Internet access, your personal website, and a few pieces of software, you, your family and friends can now keep track of your movements more easily and economically than amateur radio operators and commercial services.

As you have discovered about the amateur radio tracking program, APRS+SA, in Chapter 11, there are features that allow the program to transfer GPS data to APRS servers via the Internet. We've also seen common methods to connect your car PC to the Internet to stay in touch. APRS+SA and similar APRS and packet radio programs are not the only way to get trip tracking data from one place to another.

The Internet connectivity feature of APRS programs is intended for passing weather and other situation-specific data for special events and emergencies to and from a radio attached to your PC as a relay point – but the radio is not required – in fact the program and your PC wouldn't know and doesn't care if a radio is attached or not. The configuration and Internet capability allow you to share data directly to Internet servers just as if it was transported by radio first.

If you're not a radio geek, or want a less complicated way to share your trips with friends and family, I've figured out how to publish your travels and have them viewed on the Internet using the features of two programs – GEtrax to record your trip, and the GoogleEarth program to present maps and aerial photos of your location and routes.

Tracking people in motion over the Internet involves periodic updates of GPS data to mapping software, just like you would see locally using Microsoft Streets and Trips. The trick is getting the data from the 'trackee' (person or vehicle in motion) to the tracker – your family and friends. The more frequent the GPS location updates, the more precise the tracking. With APRS and 'Smart Beaconing' the location updates, are sent about every 20 seconds. For casual personal use, especially on long trips, updating your position every 5–10 minutes is probably more than adequate.

The process is pretty simple – first we use the GEtrax program to gather data from a GPS receiver, which it cumulatively saves to GoogleEarth-formatted data files on the hard drive of your car-PC. The 'Google file' is the key – it is the record of your location and travels.

Next, the 'Google file' needs to be transferred to your family and friends – which can be done by e-mail, or copied by FTP to a server that your friends can access to get the file. For this step we need an Internet connection (cellular or WiFi) and an e-mail program, or use the FTP program that comes with your operating system. With a few DOS commands and FTP files the 'Google file' is on its way to a server to be picked up by your family and friends. They need only access to the Internet and a copy of the GoogleEarth program to view your trip file and map your location and travels.

Putting this project together will take you about two hours – most of that time is consumed downloading and installing the software.

Parts List

- GPS receiver – one that can be configured for standard NMEA output data.

- GEtrax software from: http://web.295.ca/ ~ gpz550//GEtrax/, or Earth Bridge from: http://mboffin.com/earthbridge/

- Mobile Internet connection – WiFi or cellular data.

- Google Earth software from: http://earth.google.com (for both the car-PC to be tracked, and the people who will be watching your trip).

- Novell NetDrive from http://www.loyola. edu/5555/netdrive/applications/netdrive.exe

- The SLEEP.EXE program from the Microsoft Windows 2003 Resource Kit – from: http://www.microsoft. com/technet/downloads/winsrvr/tools/ default.mspx

- Optionally, Franson GPSGate software – from: http://www.franson.com/ gpsgate

- An FTP server to accept your files – a free GeoCities web page or one provided by your home ISP for your personal home page will work fine.

- Command and data files – provided in text below.

Steps

1. Ensure that you have a working cellular access card or WiFi connection to the Internet from the car-PC. This connection does not need to be active all of the time, but connecting more often to send the 'Google file' to your server will provide people tracking your trip with a more up-to-date view of your progress.

2. Download the software listed above.

3. Arrange for your web-site and FTP access if you are going to share your trips files with others dynamically rather than sending them by e-mail. Remember the exact settings you need to log on to your FTP site to configure

the data transfer as shown below to allow the files to be uploaded properly.

4. Install the Google Earth program on your car-PC and tell the people you want to track you to do the same.

5. Install the Windows 2003 Resource kit files to the default C:\Program Files\Windows Resource Kits\Tools folder.

6. Copy the SLEEP.EXE program file from C:\Program Files\Windows Resource Kits\Tools folder to the folder you want Earth Bridge to save the 'Google files' to – I use simply C:\NAV.

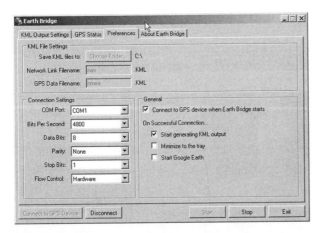

Figure 12-1 *Data file and GPS port settings appear on the Earth Bridge preferences tab.*

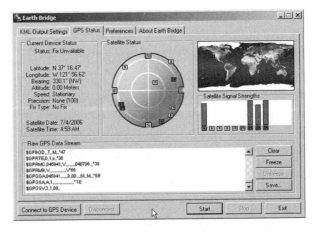

Figure 12-2 *GPS and satellite reception status reported by Earth Bridge.*

7. Install, then run and configure and verify the operation of Earth Bridge, as follows and shown in Figures 12-1, 12-2 and 12-3.

The first thing you need Earth Bridge to do is collect data from your GPS receiver and store it in a Google Earth KML file, both of which are configured in the Preferences tab of the program, Figure 12-1. To make it easy to find the Google Earth KML files I save them in a C:\NAV folder. You can name the files anything you like, but must change the data upload file described in the next step so that you transfer the correct file to your FTP/web site.

When you've configured Earth Bridge, click on the Connect to GPS Device button to enable data transfer from your GPS receiver into the program. This button is available on all of the program's screens, including the GPS Status tab, Figure 12-2. GPS Status shows you the satellites that are 'visible' to your GPS receiver, relative signal strength, and the quality of the data received as well as your coordinates.

Once you have verified that Earth Bridge is acquiring date from your GPS, you can configure how your location will appear in the Google Earth program using the KML Output Settings tab, Figure 12-3. You can give your position a title, the view angle you want those watching your location to see, and if you want the viewer to see the map as if flying into your location from outer space.

This last feature is impressive the first few times you run Google Earth, but gets rather annoying as you try to follow a trip in progress.

Be sure to click the Start button to begin saving your location and tracking data to the KML file you set in the Preferences tab.

8. Using the DOS EDIT or Windows NOTEPAD program create the following three custom command and data files in the C:\NAV folder:

Contents of the SCHEDNAV.CMD file:

```
:START-LOOP
call upload.cmd
SLEEP 300
goto START-LOOP
```

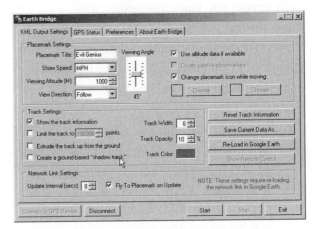

Figure 12-3 *The main Earth Bridge configuration tab to set values for the mapping display in Google Earth.*

This command file first uploads your 'Google file' data to your FTP site using the UPLOAD.CMD file, then runs the SLEEP program to wait 300 seconds (5 minutes) before repeating the file upload. While the upload process is 'sleeping' the Earth Bridge program is still saving progressive data points, but the people tracking you will receive updated 'chunks' of your trip's detailed progress.

Contents of the UPLOAD.CMD file:

```
ftp -s:upload.dat -d -v
ftp.yourftpsite.net
```

This command file uploads your 'Google file' to your FTP site using the FTP client program included with Windows, using the FTP commands and information provided in the UPLOAD.DAT file.

Contents of the UPLOAD.DAT file:

```
ftplogin@yourftpsite.net
ftppassword
put nmea.kml nmea.kml
quit
```

The UPLOAD.DAT file contains the commands and information the FTP client program needs to automatically (without keyboard input) send your 'Google file' to your FTP site. I save my 'Google file' as NMEA.KML for simplicity, and that is the file to be used by Google Earth at the viewing end. You can change this filename in the Earth Bridge program and modify this data file accordingly, allowing you to record and save different trips.

Of course, you must change the 'ftp.yourftpsite.net' (in UPLOAD.CMD), the 'ftplogin@yourftpsite.net' and 'ftppassword' items in both UPLOAD.DAT to the proper URL and logon name and password you need to access your FTP site.

9. When you have the command files created, first run the UPLOAD.CMD file and watch the display window for any error messages or hopefully no error message indicating a successful data transfer to your FTP site. When UPLOAD.CMD works you are ready to test the SCHEDNAV.CMD file to have the

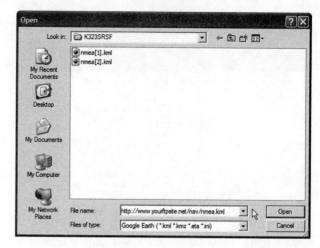

Figure 12-4 *Selecting the KML file for Google Earth to display.*

location data file transferred to your site for others to download into Google Earth. You can minimize this window to avoid distraction.

10. Run the Google Earth program on every computer that will be tracking your progress, then set the location of the (NMEA.KML) file you will be using by going to File/Open, and in the dialog, Figure 12-4, type in the web address of the file, similar to: http://www.yourftpsite.net/nav/nmea.kml then click the Open button.

11. In the Places pane of the Google Earth program, right-click the listing for the 'trip' data you just opened then select Edit. In the Edit Network Link dialog, Figure 12-5, you will be able to change some of the parameters and behavior of the data presentation to create a regularly updated tracking map. Select Refresh Parameters then under Time-Based Refresh select a time interval no shorten than the interval the data files are uploaded from the tracked car-PC. Do not set a View-Based Refresh and make sure Fly to View on Refresh is not selected. Click the OK button to close the dialog.

The results of this project will be visible on an amazing aerial-photo display fed from the Google

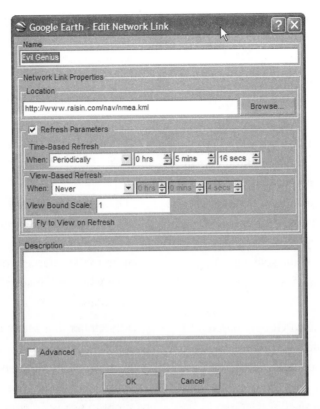

Figure 12-5   *Changing the Google Earth parameters to display an updated track.*

Figure 12-6   *The Google Earth program showing the location of the GPS/car-PC you are tracking.*

Earth servers, showing the tracked car-PC location and surrounding details, Figure 12-6.

Tracking Methods and Tips

There are aspects of this tracking project that differ significantly from the APRS-based tracking in Chapter 11 – most notable are the requirements of an Internet connection to get your data from the car to the tracker/viewer and an intermediate server to transfer the tracking data to and from, and a certain amount of inaccuracy, or perhaps lack of elegance (Smart Beaconing) in the actual tracking – something APRS has that most simple mapping processes do not.

There are also more involved Google-based tracking systems that require more complicated car-PC configurations, Linux-based servers and software, and programming with Google tools to use Google on-line maps instead of the Google Earth program. These are not necessarily easy, inexpensive or better – just different.

The Earth Bridge program is undergoing continuous improvement and may include built-in upload features in the future. A similar program, GEtrax, provides similar features in that it will create a Google Earth KML file that you must transfer to the Google Earth program through intermediate means. This 'Google Tracker' may not be elegant or smart, but using Google Earth provides some spectacular displays.

The Internet connection possibilities are many – from periodically using a cellular data modem or your cell phone to WiFi connections to a simple dial-up to the Internet you can get your data to the people you want to see it – FTP, e-mail, or your own version of uploads and web page coding.

Remember that Internet data connections using a cell phone can be very expensive unless you subscribe to an unlimited data plan (about $80 per month), but they are fairly reliable in most areas. If you opt to use WiFi connections you will not be able to upload data on-the-move as WiFi hotspots are not consistently located and work best when stationary.

Summary

I have to conclude that this is a 'pretty cool project' from start to finish. It is not difficult to implement, but it took a long time to find all of the right pieces that did not involve hours of writing or rewriting software, building, and connecting a lot of hardware bits, generally making it harder than it needed to be. It is only a matter of time before someone playing with the Google mapping software or other tools combines this into a nice tidy web service – hopefully free or at least reasonably priced – for the average PC user on-the-go.

Chapter 13

PC Weather Station

I've lived in every environment in the United States from Midwest blizzards and tornadoes to Eastern seaboard snow storms to tropical Gulf of Mexico hurricanes to the not-quite perfect West Coast. Anywhere you live, the weather is an important factor of life, work, recreation, and chores around the house.

The weather can dictate what we wear, yard work, the amount and type of emergency supplies we need to keep on-hand, and when we should head for cover from mild or severe elements. In many areas it is not enough to rely on the weather maps printed in the newspaper or what the talking heads on radio and TV tell us – we need to know if and when our immediate location is impacted.

We usually will not be as lucky as Dorothy's family in the 'Wizard of Oz' able to watch a tornado come in from miles across the plains – that funnel cloud started someplace and if Dorothy could have known the twister was forming she might not have left the farm – but then we'd have a different story. Having basic facts about the weather immediate to our location can keep us from being caught off-guard when a severe thunderstorm or tornado is threatening.

Perhaps the most practical and certainly scientific project we can do with a PC is combine computer science and meteorology into a very powerful weather station. We can keep the information to ourselves, or contribute our local weather data to the National Weather Service database so scientists can better predict the weather and study the environment.

Weather-watching, as essential as it is, can also be simply fun and informative – we can record and assess things around us, how the way we feel changes when the weather changes, and become a

better judge of wind and temperature without complicated instruments.

Creating a PC-based weather station requires a couple of simple, inexpensive components, of course a PC, and an hour or two of time to assemble, connect, and install the pieces. I found all manner of moderate-to-expensive weather station systems from Davis Instruments (www.davisnet.com), LaCrosse Technology (www.lacrossetechnology.com), Oregon Scientfic (www.oregonscientific.com), and others, but for a quick, economical start I was glad to find the 1-Wire® Weather Instrument Kit V3.0 from AAG Electronica (www.aagelectronica.com).

For a mere $79 plus import and shipping fees (about $120 total), the 1-Wire device provides wind speed, wind direction, and temperature measurements. With accessories (about $100 per set) you can add humidity and rain gauge measurements. The '1-wire' name is somewhat misleading but using conventional phone wire instead of bulky complex cabling or wireless technologies that can interfere with other things

Figure 13-1 *The heart of the AAG weather station is a set of three sensors for detecting wind speed, wind direction and temperature.*

has significant advantages – it's inexpensive, easy to work with, and has all the software you need to setup and test the system, as well as share your weather with others over the Internet. Order the AAG unit and in a couple of days you're ready to collect your live weather conditions.

Parts List

- AAG 1-Wire weather station and signal adapter ($120) from www.aagelectronica.com
- Weather Engine 5, TAI8515 test, and adapter test software, free from the AAG web site download area.
- Citizen Weather Observer Program (www.findu.com/citizenweather/cw_form.html) and Weather Underground (www.wunderground.com) registrations (both free).
- Internet connection.
- TV antenna U-bolt for weather station mount.
- 25, 50, or 100 foot telephone cable (modified).
- Crimp-on RJ-11 phone connections.

Tool List

- Common hand tools – screwdriver, adjustable wrench.
- RJ-11 phone connector crimping tool.
- Drill motor and 1/4-inch drill bit.

Warnings

Although building this project is generally a safe endeavor you do need to drill a couple of holes in the mounting arm, and for proper wind direction and speed indications you should mount the weather sensor at least 16 feet above ground – or a few feet above roof level. Proper care and safety precautions should be taken when using power tools, climbing ladders and working on the roof or on any elevation.

When working with ladders, poles, wires, and mounting the sensor unit, stay as far away (generally 3 feet or more) from power and other utility lines as possible, to avoid electrical shock and damage.

During electrical storms, to avoid damage to your PC, disconnect the cable from the sensor unit at the back of your PC.

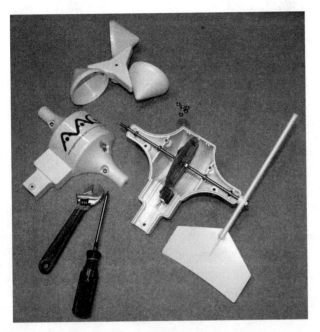

Figure 13-2 *The AAG weather station components ship unassembled.*

1. Your weather station, mounting bar, screws, adapter plug, and a short test cable come packed in a relatively small box. It is possible, Figure 13-2, to test the sensor unit, adapter, cable, and software on your PC before

Figure 13-3 *The AAG weather station assembled and ready to test.*

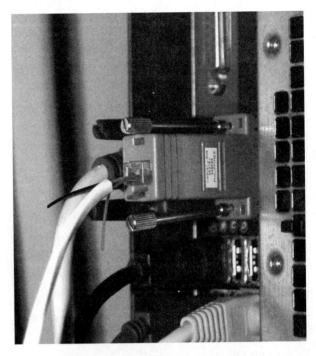

Figure 13-4 *Weather station serial/COM port adapter plug connected to PC port.*

complete assembly, but you do have to attach the wind speed cups, wind direction vane, and prepare the mounting bar, Figure 13-3, before you can mount and really use the system.

2. Download and install the AAG Weather Engine and test software. Connect the signal adapter to your PC serial/COM or USB port, Figure 13-4.

3. Connect the short test cable between the PC adapter and the weather sensor. This allows you to configure the software and test the sensor before mounting. Select the COM port used on the PC, then initialize and search for the sensors. Apply then exit the Wizzard 1-Wire Configuration when the sensor devices appear on the screen as shown in Figure 13-5. After the device configuration is complete you can spin the wind speed and direction vanes to see how the program reacts and displays.

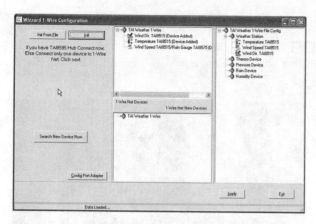

Figure 13-5 *Weather station software configuration.*

Figure 13-6 *Modifying standard phone cable to work with the weather station.*

4. While the weather station uses standard phone cable and connectors between sensor and PC, the signals are such that they can easily get lost in the cable, especially a long (25 feet or more) cable. If you are going to place the sensor unit on a 16-foot tall pole, or your roof, the cable length needed will probably be at least 25 feet.

To avoid the loss of signal (which appears as 'no device detected' in the test programs) you need to modify the standard phone cable so it can pass the data signals over a long distance. The trick is that normal phone cables use two wires parallel to each other, which creates a lot of capacitance and diminishes that sensor data signal. To avoid this capacitance we need to use nonadjacent wires.

The necessary cable modification consists of removing the original connectors from a premade cable and reattaching connectors using different internal wires, or making your own special cable. It's very easy to do this modification using a $6–10 RJ-11 connector crimping tool and a pair of RJ-11 modular plugs.

First, cut off and discard both of the original connectors on the phone cable you have. Next, using the insulation stripping blade on the crimping tool, remove the insulation from both ends of the cable. This will expose a black, red, green, and yellow insulated wire.

In a normal phone system the middle, adjacent, contacts, normally using the red and green wires,

carry the signal, and the same is true for the 1-wire weather sensor – the two middle contacts are the signal connections. We need to pick two nonadjacent wires – either black and green or red and yellow, and connect them to the middle two contacts in the connectors at both ends of the cable (we do not need the other two wires to connect to anything.)

I picked the red and yellow wires, cut off the green and black wires, insert red and yellow into the same middle two contact pins in the connectors at each end then crimp firmly. When you do this, be sure the contacts used for the red and yellow are the same at each end. Reversing them will cause the connection to fail.

When the connectors are attached you may repeat Step 3 to verify that the connections are correct and the cable works. When successful you are ready to prepare the mounting and install the sensor unit.

5. The mounting bar for the weather sensor comes completely unprepared for mounting to an antenna mast, pole, or other surface to get it into adequate position for good wind measurements. This leaves the choice of mounting up to you, though you will find that an existing rooftop TV antenna mast, and a 5–10 foot mast pole attached to a vent pipe, will give you the best results.

In my case I happen to have three antenna masts atop my office/garage/shop/Evil-Genius-lab roof to choose from. All homes have at least one vent pipe

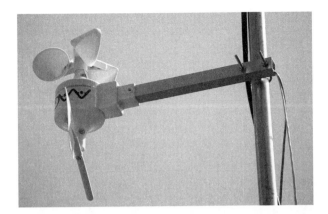

Figure 13-7 *The AAG weather station sensor mounted on an antenna mast.*

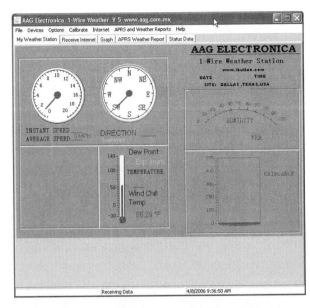

Figure 13-8 *A typical screen from the AAG weather station control program.*

for the plumbing and waste systems which make for a solid mounting option for a short mast. If you live in an apartment, your options are limited and you will likely not obtain accurate wind direction or speed readings along the side of a building.

Depending on your mounting arrangements, you will need to drill mounting holes to accommodate either a U-bolt clamp to a 1-1/4-inch or larger TV antenna mast or similar vertical pipe, or some other bolting/bracket arrangement to suit your situation. TV mast U-bolts are readily available at Radio Shack and many electronics stores.

When drilling holes in the sensor unit mounting bar, observe safety precautions, make a hole just large enough to suit the bolts but no larger so the mounting does not shift in the winds.

Check your work, run the data cable through the mounting bar and connect it to the sensor unit before inserting the U-bolt or other hardware into the bar. Mount and secure the sensor unit in position, Figure 13-7, noting the alignment of the unit with true North (to provide a more accurate indication of wind direction.)

Once mounted, retest the software and sensor configuration to be sure they work correctly.

6. When your mounted weather station is collecting and displaying data properly on the 1-Wire Weather Engine program, Figure 13-8, you are ready to share your data with the world. Register for either or both the Weather

Underground and CWOP services then configure the 'APRS and Weather Reports' options in the 1-Wire weather station program, Figures 13-9 and 13-10.

7. With your new weather station configured to share data over the Internet you can observe the records of your local weather on each of the weather reporting sites – Weather Underground, Figure 13-11, or FindU.com,

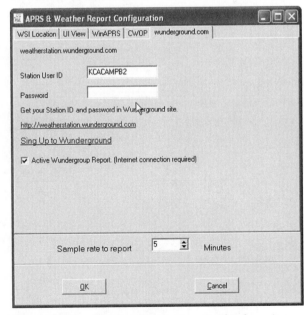

Figure 13-9 *Weather Underground upload service configuration in the Weather Engine program.*

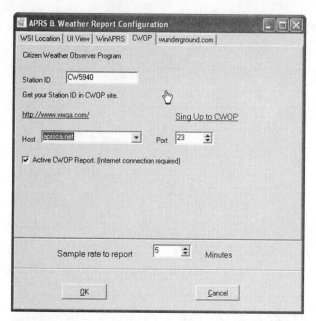

Figure 13-10 *CWOP upload service configuration in the Weather Engine program.*

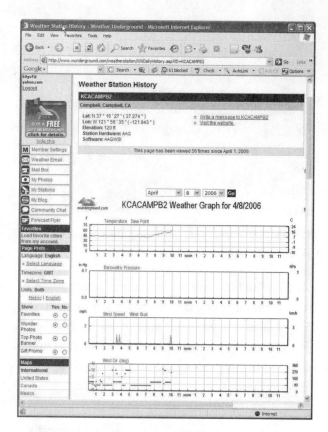

Figure 13-11 *A sample page from the www.wunderground.com weather reporting site showing data from our "Evil Genius" weather station.*

Figure 13-12. You will notice that both sites present your location in latitude and longitude and can present maps of your location.

If the location data map available at FindU.com looks familiar, it will be no surprise that the CWOP data presentation and that of APRS tracking signals is processed in much the same way. As you explore these and similar programs you will find that you can share valuable weather information as well as accurate GPS-based position information to the same Internet services, allowing you to track and share storm data for the benefit of others.

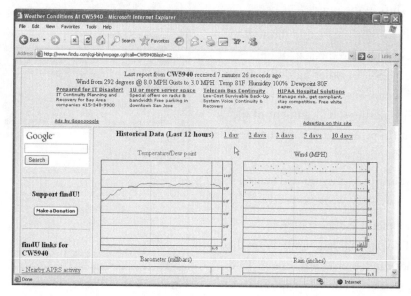

Figure 13-12 *Sample weather station data provided through the CWOP data transfer to www.findu.com.*

Summary

Of the many projects I've done with PCs, this is one of the most interesting and valuable, a notch above the trip-tracking projects in real-world applications of now everyday technology. As citizens and students of the world around us, the effects and commonality of weather in various locations is the most binding among the Earth's population. Studying earthquakes, volcanoes, and tsunamis around the world is a step above the weather, but unfortunately way beyond the present and convenient capabilities for most of us to build a project around.

Further projects

Remote Control

Virtual PC/VMWare

Chapter 14

I Want My PC TV

Playing music CDs to MP3s to DVDs proves that a PC can truly be a multi-media device – the more media the better – no matter how you download it. While we've been 'downloading' programming for years over the airwaves and cable to television sets, we've captured very little of the content, even to VCR tape, until the TiVo and ReplayTV Personal Video Recorder (PVR) appliances came along.

If you spend as much time near a PC as I do, there isn't time to play with a TiVo, much less dedicate full attention to what comes out of it, but with a high-resolution screen in front of me it makes sense to allocate some of its space to a more entertaining visual distraction like 'CSI' or 'Food TV' (come on, even geeks have to eat!).

Yes, streaming video off the 'Net is one way to get a little of the outside world into your computer time, but you cannot get full shows, and as good as broadband can be, it seems a shame to waste it on choppy 'TV'. An alternative is to add another multimedia input to your PC and bring full quality TV to the screen with a TV tuner card.

TV tuner cards, like the Pinnacle Systems PCTV card, Figure 14-1, have been available for quite some time but only recently have they been cable-ready and PCs become fast enough to watch without glitches and record a show at the same time – creating your own PVR without taking up any more space with more appliances and gadgets.

Many cards also provide composite and S-video inputs so you can connect a cable or satellite input, VCR, DVD or PVR as optional sources of video content, Figure 14-2.

Setting up a TV tuner on your PC will take less than an hour of your time, cost under $100, and keep you entertained for hours during your surfing, email, and downloads, and let you schedule and record shows for later viewing.

Figure 14-1　*A typical PC tuner and video capture I/O card.*

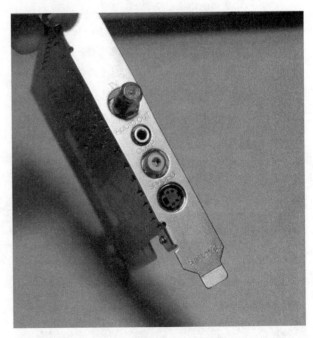

Figure 14-2　*Tuner cards may also support composite and S-video sources.*

Parts List

- Your PC with an available PCI slot.
- TV Tuner card – there are many brands to choose from including ATI, Pinnacle, and Hauppauge.
- Hard drive upgrade to accommodate saving large video files.

Steps

1. This may be obvious, but not following it may cause your installation to fail: read the instructions with your tuner card to determine if you must install the driver and card operating software *before* installing the card. If you install the card first, Windows may not recognize the new hardware and lock it down to being an 'Unknown Device' requiring correction before you can proceed.

2. Safety first and always. Shutdown your PC and disconnect the power cable.

3. Open the case of your PC and locate an unused PCI card slot. Install and secure the tuner card in the slot, Figure 14-3. Close the case.

4. Connect an audio jumper cable (provided with your tuner card) from the Line In or

Figure 14-4   *Audio cable connection between PC sound card and tuner card.*

Microphone input on your sound card to the tuner card audio output jack, Figure 14-4. Connect your choice of TV antenna, cable, composite video, or S-video to the tuner card.

The composite and S-video signal inputs to the tuner card do not provide audio. Audio from composite or S-video source devices must be connected from those devices to your PC's sound card input(s) and configured as the audio source in the tuner/capture program.

Figure 14-3   *Installing a TV tuner card in an open PCI slot.*

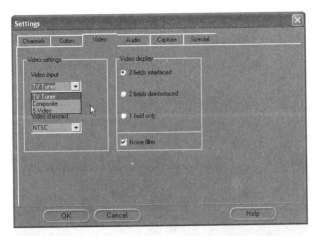

Figure 14-5 *Selecting the video source for your tuner card.*

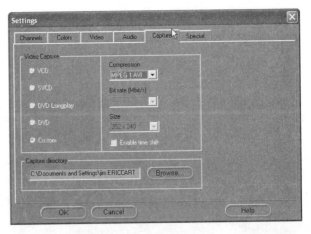

Figure 14-7 *Setting the video recording file compression type and location to save recorded shows.*

5. With your card installed and connected, reconnect the power cord to your PC and boot the system into Windows. When Windows is running, run the software for the tuner card and begin its configuration, Figure 14-5.

Since most tuner cards support normal VHF and UHF TV channels as well as several variations of cable TV systems – Comcast, Adlephia, etc. – for taking input off-the-air you can select a TV tuner and let the card do the work. If you are using an existing cable box you may be able to use composite or S-video input, but then you must use the cable box tuner and remote control to change channels.

6. Select the audio input and controls, Figure 14-6, for hearing your video source. At the **TV playback input** control select the sound card jack/source for use with the TV

tuner (not the composite or S-video input.) If you are using composite or S-video select the sound card input/jack corresponding to these sources under the **Sound recording input** control. Adjust the **Sound recording level** control so the bar graphs indicate about 80–90% across the graph area.

7. Next, configure the type and location for saving program content, Figure 14-7. Most tuner control programs support the old MPEG-1 AVI format, but may offer plug-ins and codecs for MPEG-2, and the new high-definition MPEG-4 format.

8. Before you can easily use the TV tuner card you let it scan across all available TV channels, Figure 14-8, to determine which

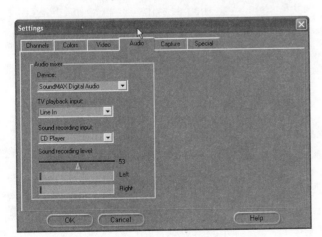

Figure 14-6 *Selecting sound inputs and volume levels.*

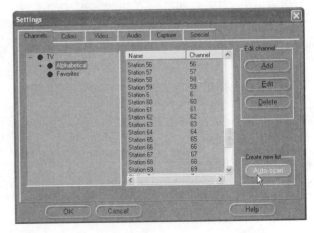

Figure 14-8 *Setting up the TV channel selections in the TV tuner configurations program.*

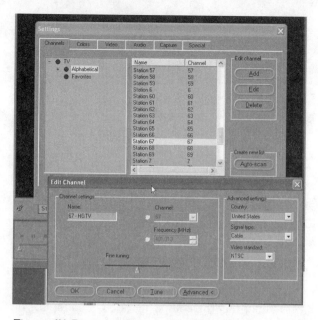

Figure 14-9 *Providing meaningful labels to your channel selections.*

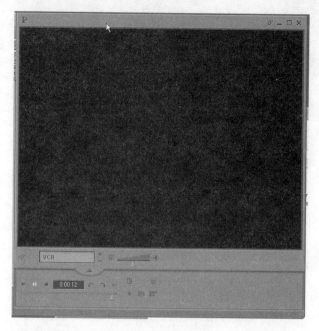

Figure 14-10 *The active PC TV viewing screen. The video display is dynamic and cannot be screen-captured.*

channels are in use rather than making you click through dead spaces.

9. Off-the-air programming does not provide any intelligence about the TV station or network that appears on a specific channel, so the tuner program allows you to edit the channel labels, Figure 14-9, so they make sense to you.

10. Using your TV card is a snap. You get a virtual TV picture tube window, Figure 14-10, and controls for channel, volume, recording, and taking a snapshot of a scene. Since the TV picture is always changing you cannot use conventional screen-capture programs to save a scene.

Summary

Short but sweet... adding a TV tuner card is one of the best ways to optimize costs and space and still enjoy a lot of live video content – from TV to movies – and save them for later viewing. The trick is getting the video source – TV antenna or

cable signal – to the back of your PC. In Chapter 15 I'll show you how to get a variety of video sources from an existing TV/multi-media setup to PCs anywhere on your home network.

Chapter 15

Sling Video All Around

A new twist on putting a TV tuner card inside your PC is to *not* put a tuner card in your PC, but instead have a tuner 'attached' to your PCs on your home network, to be shared with any PC in your house, or over the Internet.

The evil geniuses at SlingMedia created their Slingbox network appliance to stream either off-the-air TV signals, composite, or S-video input over your network. There are similar products sold by LinkSys and D-Link to share either video or audio over wireless. Slingbox is unique because it is also a server that hosts your entertainment making it available on any PC with the SlingPlayer program and an Internet connection.

The Slingbox, Figure 15-1, and SlingPlayer program can also remotely control your VCR, cable box, or DVD player so you do not have to get up from your PC to change channels on devices that you connect through S-video instead of using the built-in tuner.

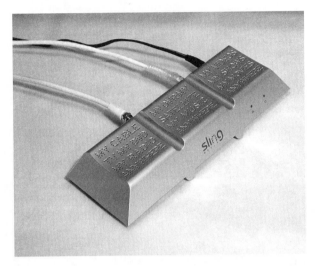

Figure 15-1 *The sling box – simple but very powerful.*

This project will take less than an hour; the Slingbox is available online and from most electronic retailers for about $200, the player software is free, and the ability to enjoy your home entertainment from just about anywhere is priceless.

Parts List

- Slingbox.
- Network connection – wired Ethernet or a wireless bridge.
- SlingPlayer software for each PC from http://www.slingmedia.com/us/support/downloads.php

Steps

1. Unpack and connect your Slingbox to your video source(s) using the cables included, or your own, and your home network,

Figure 15-2. If the device you are connecting your Slingbox to is a cable box, DVR, DVD player, etc. you should also connect the

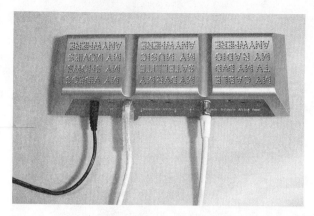

Figure 15-2 *The Slingbox needs only three simple connections – power, network, and TV signal input.*

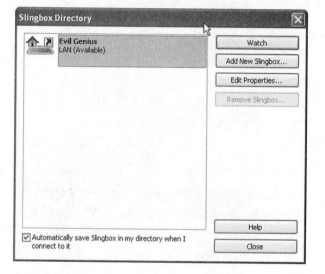

Figure 15-3a

Figure 15-3b *SlingPlayer's first installation screen.*

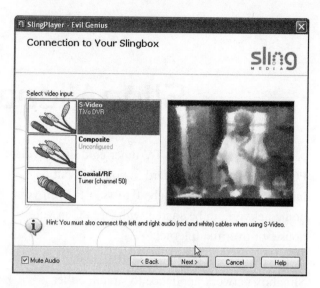

Figure 15-4 *Selecting one of three available signal inputs to the Slingbox.*

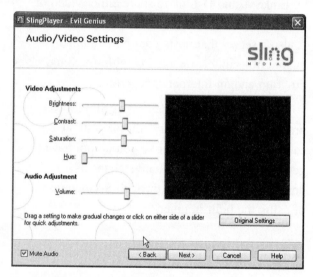

Figure 15-5 *Adjustments for Slingbox picture quality and sound level.*

included infra-red (IR) remote control cable and place one of the IR-emitters in front of the player. This is the simple and boring part.

2. Installation of the SlingPlayer software takes a bit of thought as you go through the process. The first screen, Figure 15-3, asks you to select if this is the first time you're installing your Slingbox – which you are at this point, or if your Slingbox is already setup – to be used for installing on other computers or re-installing. The second

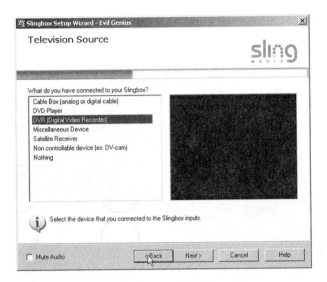

Figure 15-6 *Selecting the type of device that feeds signal to the Slingbox.*

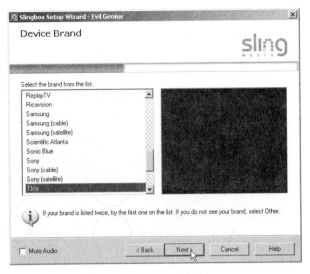

Figure 15-7 *Selecting the manufacturer of the device feeding your Slingbox.*

screen should simply discover your Slingbox on your network.

3. Since the Slingbox has three inputs for video signal – antenna (air), composite video, or S-video – you must select the signal source used in Step 1 to begin configuration, Figure 15-4. If you have only cable or an antenna, select Coaxial/RF. If you have a TiVo, cable box, DVD, or VHS player with S-video output, select S-Video for the cleanest signal. Click the Next button to continue.

4. The next setup screen, Figure 15-5, provides adjustments for picture appearance and sound level. This adjustment is perhaps a bit premature in the process, and it is difficult to tell from the minidisplay how good or poor the picture looks, so I'd simply accept the default values and adjust this after.

5. Depending on the source signal (RF, composite or S-video), setup prompts you for the type of device connected to the Slingbox, Figure 15-6. This affects the next few screens where you setup, make, model and remote control operations.

6. On the next two setup screens, Figures 15-7 and 15-8, you select the make and model of the device feeding your Sling Box. In my example I am using a TiVo Series 2 DVR.

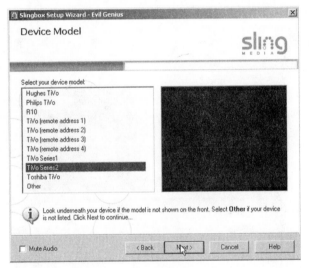

Figure 15-8 *Selecting the model of the device feeding your Slingbox.*

7. SlingPlayer setup's Channel Changing screen, Figure 15-9, lets you tinker with the settings to make the IR remote control functions work best. The default settings should be fine for known devices, but you should test the channel changing function to be sure. Beware the positioning of the IR-emitter from Step 1, it must be able to 'see' the IR-detector on the device. The double-stick tape does not fully support the weight of the emitter to keep it in place so you may have to add additional tape to secure it.

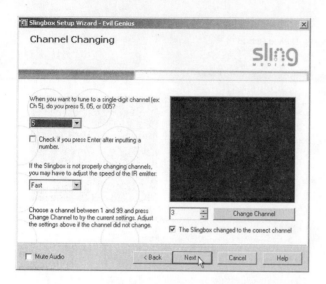

Figure 15-9 *Adjusting the remote control settings Slingbox uses to control the source of its signal.*

Figure 15-11 *Assigning passwords for access to your Slingbox.*

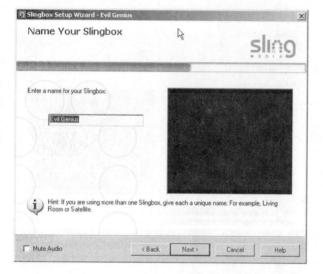

Figure 15-10 *Giving your Slingbox a friendly name.*

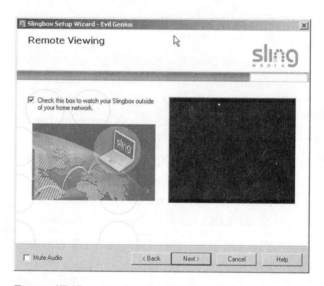

Figure 15-12 *Configuring Slingbox to stream to your Internet connection.*

8. Once you have configured the input to your Slingbox you can give it a unique name, Figure 15-10. This will help distinguish this specific unit from others on your network.

9. Once your Slingbox has a name it needs two passwords, Figure 15-11, one for regular users/viewers, the other for the owner/administrator.

10. If you want to share your Slingbox over the Internet you need to check the box, Figure 15-12, so others can watch.

11. Your Slingbox can be automatically configured over your network, Figure 15-13, but my preference is to know exactly what the configuration is to leave out the guess-work, so I select Manual (advanced) configuration, then verify the network address settings on the next screen, Figure 15-14. By default the Slingbox always takes an address ending in '237', such as 192.168.1.237 or similar – probably because that's the number of the local highway near Slingbox headquarters?

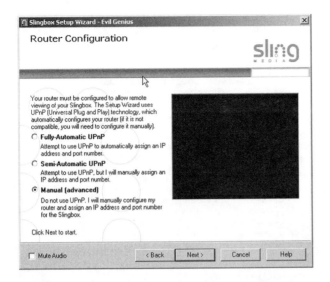

Figure 15-13 *Selecting automatic or manual network settings.*

Figure 15-15 *Making the IP port assignment Sling Player will use to access the Slingbox.*

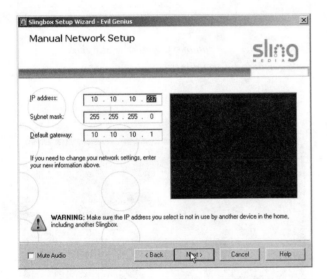

Figure 15-14 *Manually assigning TCP/IP parameters for your Slingbox.*

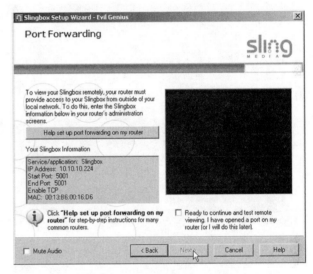

Figure 15-16 *Details you will need to configure your firewall/router for Internet access to your Slingbox.*

Note this IP address for your home firewall/router configuration.

12. When you elect to share your Slingbox it assumes you want to use IP port 5001, Figure 15-15, but you can change it to anything you want that is not used by other devices on your network, and that you can pass-through your home and other firewalls. Note this port number for your home firewall/router configuration.

13. When the network configuration is complete the setup program gives you a summary of the settings, Figure 15-16, you will need for your firewall/router.

14. The last step of setup is revealing 'Finder ID' of your Slingbox, Figure 15-17. This is a unique 'serial number' you can share with others so they can find and watch your Slingbox remotely.

15. Upon completion of setup, your Slingbox is ready to watch. Repeat the installation on other PCs as you want. Each SlingPlayer user will have access to changing video sources, channels, and display settings to

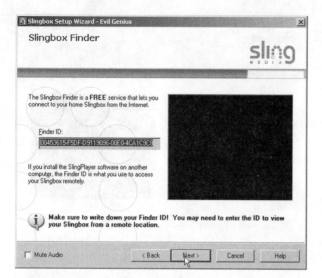

Figure 15-17 *Each Slingbox has a unique numerical 'Finder ID' to identify it.*

Figure 15-18 *The Slingbox configuration dialog with access to password, audio/video, network and firmware settings.*

suit personal needs. You can also go back into setup by selecting Slingbox, Properties, then Slingbox configuration, Figure 15-18, to configure your other video sources.

16. Should you forget the details about your Slingbox, you can always retrieve its data from Slingbox, Properties, Information screen, Figure 15-19.

Figure 15-19 *A summary of all the Slingbox configuration information.*

Figure 15-20 *The final results – watching TV with the SlingPlayer.*

When everything is configured and working, well, you'll be watching your own TV streamed over your own network, Figure 15-20.

Get Creative

Since the Slingbox can accept input from three different video sources, and by remote control switch channels on multiple sources of input, you have the ability to switch between cable TV, a home security camera, your TiVo, etc. If you have a TiVo or similar you can select additional input sources through it and expand the possibilities. You can also enjoy your Slingbox on a SlingPlayer-equipped PDA.

Summary

The Slingbox is not something you do *to* your PC, but something I think you do *for* your PC. You've just added a very powerful peripheral available to all your PCs, at home and away. In fact, you might try accessing your Slingbox as you cruise down the road with your car-PC and mobile Internet access!

Chapter 16

Voice-Over-IP With Skype

So far we've explored almost everything-Internet from mobile data to navigation to GPS-tracking to streaming video all over. Our next project is to delve into the world of voice-over-IP (VoIP) otherwise known generically as Internet-telephony. By all the press and marketing hype VoIP would seem to be a new technology just appearing on the horizon for computer hobbyists. In reality VoIP has been in use for many years as the backbone of most local and long-distance telephone service providers and satellite broadcasting.

While networking giants Cisco and Northern Telecom were creating VoIP phone systems for corporations, once again amateur radio operators were tinkering with VoIP or digitized voice on many fronts, from audio-to-digital modem attachments to radio equipment to PC sound card interfaces and software to connect ordinary two-way radios to PCs and 'broadcast' over the Internet.

The later efforts created two significant Internet-based radio interconnection networks – the Internet Radio Linking Project (IRLP) (www.irlp.net) and EchoLink (www.echolink.org). Both of these allow amateur radio stations worldwide to bypass the vague characteristics of RF propagation and tie radio systems together through PCs over the Internet. Thousands of radio operators around the globe now enjoy Internet-linked radio chatter made possible with a couple of PCs, a bit of software, and an easy bit of wiring – in fact the same bit of wiring that allows you to do packet-radio data transfers to send messages or APRS position data over the air to Internet servers – yet another reason to get your amateur radio license!

While the telecom giants toiled on conventional money-making opportunities and the amateur radio community evolved its methods for VoIP, the instant messaging folks at AOL, MSN, and Yahoo! added voice then video chat capabilities to their services, while new 'dot-com' companies like Vonage and Skype sprang up to create a consumer VoIP marketplace. Adding to the fray, the makers of home routers, LinkSys, D-Link, and NetGear, are selling VoIP appliances that take the requirement for having a PC or Mac out of the Internet phone service business.

Soon, if not already, the names 'Vonage', and 'Skype' will become household words among the techno-savvy home computer users. Skype is perhaps the easiest service to get involved with because you probably already have what it takes to 'VoIP with Skype', so let's get calling shall we?

Parts List

- PC with sound card.
- Broadband Internet connection.
- Speakers and microphone, or headset.
- Skype software and sign-up – www.skype.com

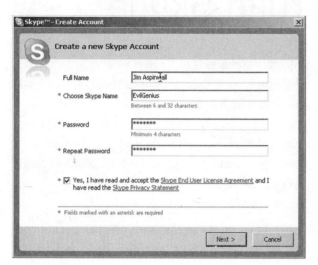

Figure 16-1 *The first Skype setup screen lets you establish your Skype name.*

PC Mods for the Evil Genius

There is essentially one extended 'step' to this part of the project – download the latest version of the Skype software from www.skype.com, then install it. It takes about three minutes to download and another 2–3 minutes to install and setup through the next three screens:

As with many other instant messaging programs, Skype offers you options to select for your online status (Busy, Away, Offline, etc.), adding contacts.

Figure 16-2 *Skype setup needs your email address to keep in touch should you lose your password.*

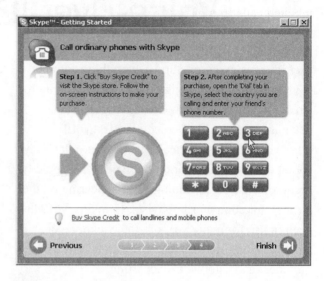

Figure 16-3 *The completion of Skype setup, and a mild nag to subscribe to regular phones and get your own Skype phone number.*

Figure 16-4 *A Skype VoIP 'phone' call in progress to another Skype user.*

Skype-to-Skype calls are but one way to use the service. For a fee you can make Skype PC-to-landline calls and for another you can get your own landline phone number so you can take inbound calls from nonSkype users. Skype may not yet replace your regular phone line, and when you read *VoIP is Not Emergency-Ready* later in this chapter you will understand why you might not want to replace your Plain-Old-Telephone, but it and similar services are improving performance and services to be more like regular phone service.

A Phone Patch for VoIP

With services like Skype it is obvious that someone would create a telephone peripheral to take the place of the speaker and microphone or headset attached to the PC. Checkout almost any electronics or computer store, or online retailers like www.CyberGuys.com and you'll find a dozen different flavors of 'VoIP phones' that are not self-contained phones per se but USB 'sound cards' built into a phone-like handset gadget. Some of these gadgets will work with all VoIP services, and some with only selected services the manufacturer has partnered with. The cost of these 'phones' ranges from about $30 for a wired device to $300 for a WiFi version you can use at any hotspot – no PC needed.

In true Evil Genius-style, you may find it more satisfying to mod-up your favorite conventional phone – corded or cordless – to use with Skype. For this part of the project you'll build a simple interface between a conventional phone and the connections for your PC's sound card.

This project should take about an hour of your time, more if you assemble it into a nice project box and make it 'nice looking', and cost you less than $20. You will need some tools for wiring and soldering, and a handful of parts available at real electronics stores or from Jameco, DigiKey, or Mouser parts houses online.

Parts List

- Conventional TouchTone™ phone (look in the junk box or garage).
- RJ-11 jack (available from Radio Shack or similar).
- TRIAD MAGNETICS #TY-145P transformer, www.Jameco.com catalog #630459.
- 9 volt battery.
- 9 volt battery connector with wires.
- 4 each 330 ohm 1/4- or 1/2-watt resistors (any value from 150 to 470 ohms will work fine).
- 6 foot 1/8-inch stereo plug patch cable (cut in half).

The adapter, coupler, or interface you will be building, Figure 16-5, has two functions – the first is to provide a small amount of voltage to make your conventional phone work, the second is to couple audio signals between your phone and your PC's speaker, and Line In or Microphone connections. Amateur radio operators and people who work with phone circuits will recognize this circuit as that of a typical phone patch or hybrid interface commonly used to interface telephone lines to radio or broadcast equipment.

The first function, making the telephone work, is provided through the RJ-11 phone jack, with the red and green wires connected to one side of an audio transformer and a current source – the 9 volt battery. A conventional phone consumes as little power as a small transistor radio so the 9 volt battery is perfect to provide a power source for the phone. With a little modification this is almost exactly how every analog telephone in the country gets its power.

The second function, coupling audio between the phone and PC, is provided by the transformer, resistors, and the stereo plugs connected to the PC sound card. Because the connections from the transformer to the PC are the same, the stereo plug connections to the sound card can go to either the Speaker Output (Line Out may not have enough volume to be heard adequately in the phone) or the

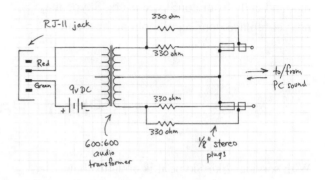

Figure 16-5 *Schematic diagram of the Evil Genius VoIP phone patch.*

Line In or Microphone input. The 'magic' of this side of the circuit is that the PC connections to the transformer cancel each other out, with the common or ground line tied to the center connection of the transformer, so you should not experience feedback or echo during use.

Building this circuit requires no special skills or techniques except careful soldering of components. One 1/8-inch stereo plug-to-plug cable is cut in half to make each of the PC connections. Either side of the transformer may be used for either end of the interface.

After construction, plug everything in, make a test call over Skype or your favorite VoIP application, adjust your volume and microphone levels, and talk away.

VoIP is Not Emergency-Ready

As a disaster services worker and instructor, involved with emergency communications under all sorts of adverse conditions, I feel it is important to give you a heads-up about some of the realities

and concerns over Internet-based and wireless phone services in general.

Before you give up your conventional land-line phone service for VoIP over cable or DSL-only

Internet access, you should know that there may be some serious drawbacks that could affect your ability to quickly obtain emergency police, fire, and EMS services when needed.

One critical issue is with caller-ID related to the location of the caller. With typical landline services the phone company transmits exact address information to the 9-1-1 emergency phone centers, and your local police, fire and ambulance services. This is because your hard-wired analog phone service is established at a fixed, known address. This is not the case with VoIP services – while the IP address dished out by your ISP can identify your current Internet connection, it is not tied to a specific location. This means that even if you live in North Platte, Nebraska, if you use your Vonage or Skype phone service from a hotel in Bethesda Maryland – no one will know where you are.

Only in a few areas, and not all services, can you get location services fed to your local 9-1-1 emergency dispatch center to enable quicker, more accurate response. This is similar to the presence or lack of Enhanced or E-9-1-1 services from your cell phone, where the cellular phone system and a GPS receiver in your cell phone combine to provide somewhat accurate location data when you are on the move. Location-based services are not the only caveat with Internet-based voice calling services.

Another concern should be readily apparent – what if the power goes out? Unless your home digital equipment, DSL modem, PC or VoIP appliance AND your ISP are equipped with backup power systems, you can kiss your digital-world goodbye during most disasters – thunderstorms, ice-storms, earthquakes, floods, and the nondisaster of a rolling brown- or black-out due to high energy consumption or routine power utility maintenance.

Summary

Technologies like Voice-over-IP and Skype applications are a tremendous extension of the Internet, personal computing, and communicating with others. Add a webcam and you've got the videophone MaBell could only hope for 20, 30, even 40 long years ago when such a thing was first demonstrated to the public.

Chapter 17

Security Webcam

While a webcam can be a nifty gadget for a select few online chats, I don't think anyone really wants to spend time watching me type and stare at my displays, sip wine, puff a cigar … well, you get the idea.

Unless you get to spend all day at home, what's your webcam watching while you're away? Nothing!? Why not put it to good use letting it watch your cat, the nanny, cleaning lady, grass grow, catch that annoying neighbor tossing things into your front lawn, or simply keep an eye on your valuables?

Obviously just leaving your webcam on isn't going to do much for you unless the software that operates it can broadcast or capture some specific activity of interest, and preferably share it to a web-site or send you an email if something happens.

It just so happens that there are quite a few webcam security programs floating around that can put your camera to good use when you're not the visual subject of interest. Out of the many choices the one I like best is called 'webcamXP' – from www.webcamxp.com It offers a decent trial period to get used to the features and a reasonable price for either the personal version to simply stream up to 10 video sources, or the 'pro' version that can handle 50 video sources and includes motion

Figure 17-1 *Is your webcam sleeping? This Logitech camera can keep watch on your stuff.*

detector and alerting features. As a novelty to simply watch your grass grow, the personal version is more than adequate, but for the enhanced security features the 'pro' version is a must.

Starting fresh, this is not an inexpensive project – you need to have a 'webcam' which will cost anywhere from $30–80, and the software which is either $40 or $80. Considering the potential value of catching the cleaning lady rifling through your jewelry box or snacking on your truffles, the $70 to $160 expense may pay for itself in no time. With all the goods at hand it will take you 30 minutes or less to get the project going, though you'll probably play with it for hours after.

Parts List

- Webcam – a Logitech QuickCam or similar.
- Your PC.
- Internet connection.

- WebcamXP software.
- A web-host to upload your camera shots for sharing.

1. The first thing is to have a webcam, and the second, downloading and installing your preferred version of webcamXP. There are quite a few options, Figure 17-2, to the program's setup, but at first-use 'it should just work' as shown in Figure 17-2.

2. Setup your personal website or Internet hosting to accept a known FTP file transfer (username and password) from the webcamXP program. Once you've tested and verified that this works, put the logon parameters into the webcamXP configuration, Figure 17-4.

WebcamXP also provides its own web-server so you can access the images and interactively chat with the host system on your local network or over the Internet if your ISP allows in-bound server connections.

3. Not that you need to change anything, but webcamXP offers a variety of options, Figure 17-5, for recording what your camera sees, from the filename to the video and audio encoding formats.

4. The security features of webcamXP include a visual motion detector that becomes the 'alarm system'. When sufficient motion is detected above the level you set, Figure 17-6, the program can capture a snapshot, record a video of the activity, and send email alerts to you with a picture of the 'action' attached.

5. Provided uploads to your home page website are successful you can visit the page and check on your camera from anywhere on the web, Figure 17-7.

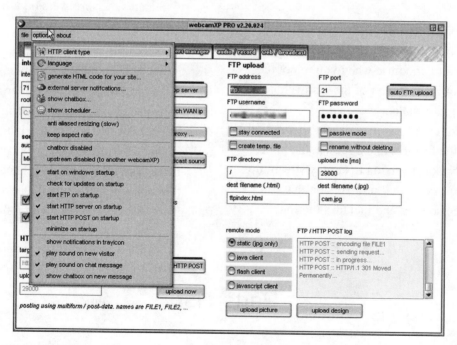

Figure 17-2   *WebcamXP's option list appears daunting but is seldom needed.*

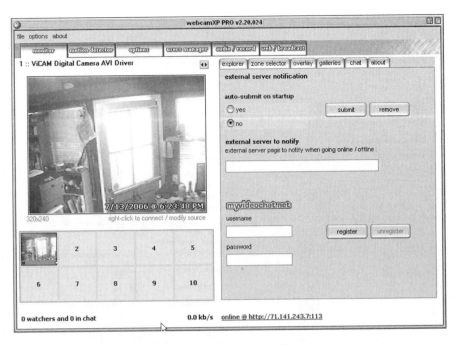

Figure 17-3 *WebcamXP will monitor and time-stamp your camera's images.*

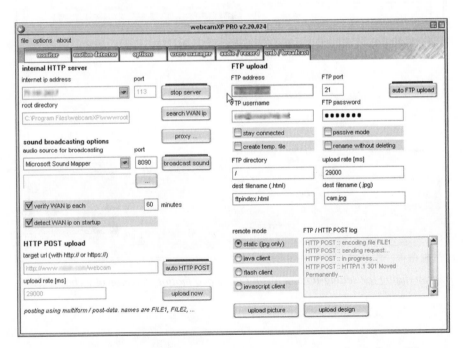

Figure 17-4 *Configuring FTP upload, web-page, and web-server parameters for webcamXP.*

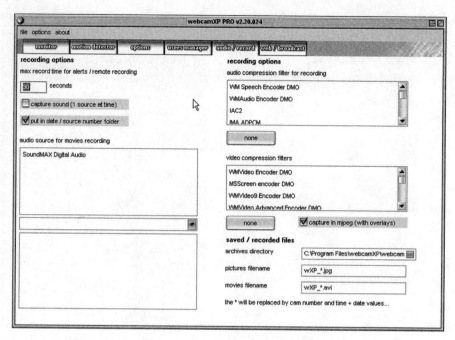

Figure 17-5 *Selecting the audio and video recording options for webcamXP.*

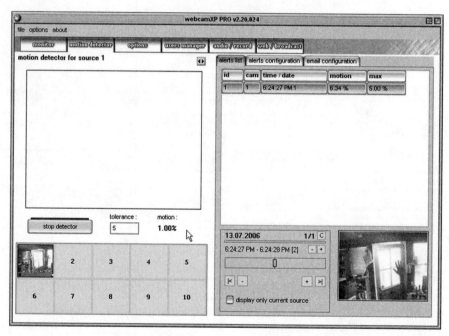

Figure 17-6 *Program dialog for setting-up and testing the video motion detector in webcamXP.*

6. For local viewing, or over the Internet if you can access your home PC through your ISP, you can interact with the camera sources and watch live streaming images, Figure 17-8, through the built-in web-server.

Figure 17-7 *Viewing a periodically updated snapshot from webcamXP to a personal web-page.*

Figure 17-8 *Live streaming images from a webcamXP server.*

Summary

There's nothing like a picture to fill in for a thousand words when the right moment is caught on camera. Although a very simple, almost 'plug and play' project to accomplish, implementing a web-cam-based security system, can provide a little entertainment or a lot of value, whether or not you also have a conventional security system.

If you live in a truly interesting location such as San Francisco, New York City, Miami, Maui, Seattle, or Ketchican Alaska it might be a real treat for the general Internet community to share your scenic views; a broadcast webcam is the way to do it. A most impressive product for simple scenic viewing is the Sony SNC-RZ30N Network Camera which offers remote control pan, tilt and zoom features through a Sony-hosted web-site.

Either way you choose to try cam-ming you'll find it a lot of fun, perhaps educational, and quite possibly invaluable to keep an eye on your stuff.

Chapter 18

Contributing to Science Explorations

Globally, science experiments consume hoards of computing power to perform millions and billions and trillions of calculations. One or even a handful of supercomputers is not enough to keep up with myriad computing-intense research projects around the world.

With approximately 200 million personal computers worldwide, as a whole, PC users have the equivalent of hundreds of super-computers just sitting under desks, on laps, on café tables, in closets, and heaped in scrap piles. Most of this computing power sits unused, with the power switch turned on, awaiting commands for something to do.

If you'd care to share a little of your CPU time, your computer could be the one to help find the pieces for the ultimate cancer or AIDS fighting drug, discover a distant life form, or crack any number of data encryption codes, while you're sleeping or getting another latte. No less than 79 public distributed computing projects, based on a list of projects at the Distributed Computing website, www.distributedcomputing.info are hungry for your CPU time.

Sharing computer CPU time is not new – it's how mainframes always worked and something Microsoft Windows, Unix and Linux, and now Mac OS X, have done for years. In fact, every time you start your PC its CPU is locally time-shared between things Windows needs to do to be Windows, and the applications programs you run. Running an additional application or two won't bother most modern PCs – whether that application shares its data over the Internet or locally.

For sharing your CPU time between e-mail, word processing, spreadsheets, the Web, and one or more science-experiments, I'd have to say a 'modern PC' at minimum means a 1.2 GHz or faster Intel Pentium III or equivalent AMD Duron or Athlon processor, at least 512 MB but preferably 1 GB of RAM, and an ATA-100, ATA-133 or SATA hard drive. If you simply want to donate the power of an old PC sitting in the corner you can contribute something less, depending on the requirement of the program you choose to run.

Each of the applications you will encounter allows you to limit the amount of CPU time, memory, and disk space it consumes, and when it is allowed to process data and connect to the Internet to submit results or get new 'assignments' to work on. Your Internet connection may be dial-up, ISDN, DSL, cable or through a proxy server at work (work permitting of course) and does not have to be active all of the time.

These on-line computations are not merely background tasks using your PC like some pack-mule with no interest or value to the user – you can actually see the progress and results of your contribution. The Folding@Home DNA and protein sequencing project provides a nice visual display of a changing protein molecule. The SETI@Home project provides a visualization of the radio spectrum analysis it is working on while Einstein@Home shows star maps as they surround the Earth. NetDIMES provides a variety of plots of traffic routing throughout the Internet.

Getting started on any one of these projects requires no special effort or skills on your part. You will not need tools, parts, wires, nor worry about drilling holes in or burning your fingers. If you can fill out an online form, download then install one or more pieces of software, you can become a 'rocket

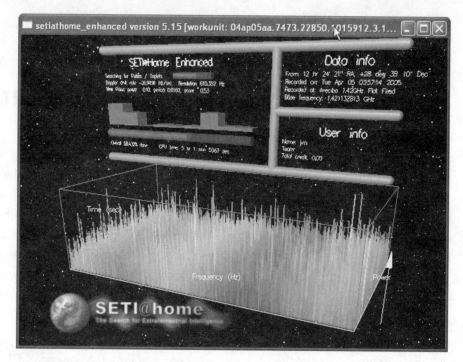

Figure 18-1   *The SETI project analyzes patterns in radio waves traveling through outer space.*

scientist', 'molecular biologist', 'astronomer', or 'code cracker' within a few minutes.

Step One – pick a project, or several. From the electronic search for signs of life from other galaxies, to mapping human genomes, to global climate studies, there is a project of interest to just about everyone.

The Search for Extra-Terrestrial Intelligence (SETI) project, Figure 18-1, is perhaps the most well-known and the most popular project available. A collaborative effort between the scientists at SETI.ORG and researchers at the University of California at Berkeley, SETI@Home is managed online through the Berkeley Open Infrastructure for Network Computing (BOINC) resources – http://boinc.berkeley.edu

Einstein@Home, Figure 18-2, also managed through BOINC, struck a familiar note with me because it's run out of my home state by the University of Wisconsin – you could say I'm giving equal (CPU) time to THE Berkeley and 'the Berkeley of the Midwest'.

Speaking of equal time, Stanford University (similar colors to the U of W) gets some CPU cycles on one of my PCs for their Folding@Home project, Figure 18-3, which has given my PC a homework assignment of working on the 'p1809_Collagen_POC AMBER core' molecule – whatever that is. At the going rate of 18 minutes per portion of the 'assignment' of 500 portions, I think this chapter will be finished before my science homework is completed.

Another project which I'm running is the Distributed Internet MEasurements & Simulations, (DIMES) or NetDIMES Internet mapping project, Figure 18-4. The program is apparently given an Internet address to trace from my 'homework' PC, collects a bunch of data from each hop along the way and sends it back to researchers in Israel. For all I know it could be trying to intercept messages about illegal money laundering or world-cup soccer scores.

Berkeley Open Infrastructure for Network Computing (BOINC) brokers 13 different distributed computing projects and manages them on your PC with one piece of software, the BOINC

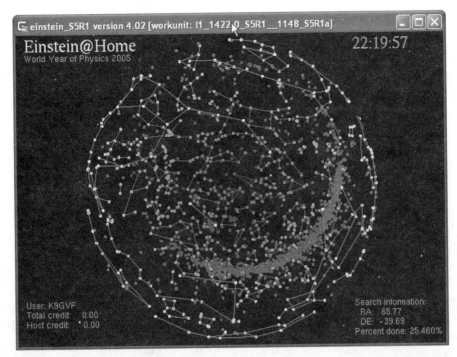

Figure 18-2   *Einstein@Home looks for pulsars by sifting through data from two gravitational wave detectors.*

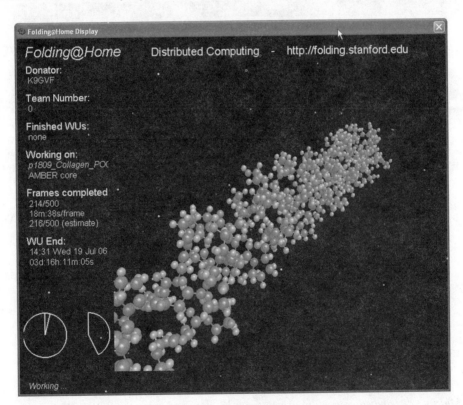

Figure 18-3   *Studying the process of normal and abnormal protein assembly or folding is the work of Stanford's Folding@Home project.*

Manager, Figure 18-5. BOINC Manager allows you to watch the progress of the projects you select and provides web links to the configuration parameters for each project.

For more information on these and myriad distributed computing science projects visit the following web-sites:

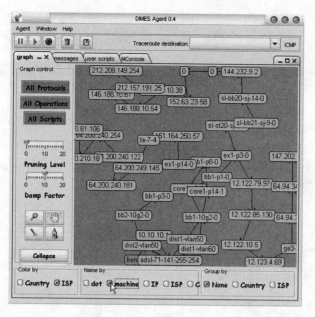

Figure 18-4   *Analyzing data paths from source to destination and back is the objective of the NetDIMES program.*

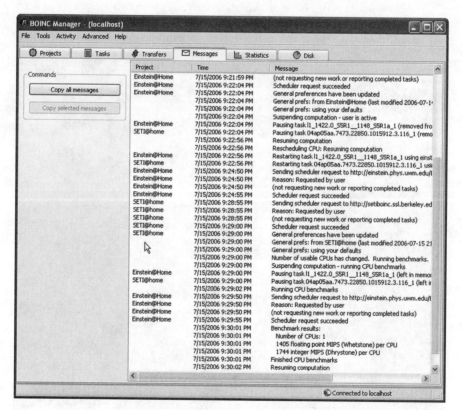

Figure 18-5   *Multiple distributed computing projects can be managed using the BOINC Manager program.*

- http://BOINC.berkeley.edu which manages the following experiments:

  - BBC Climate Change Experiment

  - Climateprediction.net

- Einstein@home

- Tanpaku

- SIMAP

- World Community Grid

- LHC@home
- Predictor@home
- Quantum Monte Carlo at Home
- Seasonal Attribution Project
- SETI@home

- Rosetta@home
- SZTAKI Desktop Grid
- http://Folding.stanford.edu
- http://www.netdimes.org
- http://distributedcomputing.info

Summary

The 'PC mods' here represent adding purpose to, or repurposing your PC to do something a little more useful than grabbing headlines from CNN or mundane tasks like paying bills and checking sports scores on Yahoo!

Almost anything you do or can do with a PC can relate to something else. Combining our earlier weather station data sharing project with any one of the global climate studies listed above can result in a comprehensive 'all-weather PC' 'mod' you and your local school can be proud to be a part of.

Running a NetDIMES session while you are traveling across country using cellular and WiFi connections may provide valuable insight into how well the Internet is put together, or not.

As you dig into the specifics for each of the available distributed computing projects you will be amazed to find out how much life-saving research a few thousand extra computers sharing massive workloads can be accomplished while you sleep, go to work, or watch DVDs in the den.

Chapter 19

Keep In Sync

Time-wise that is. The real-time clock circuit in your PC can drift faster and slower as much as 2–3 minutes per day – certainly not acceptable if you really want to be on time for appointments, schedule digital video recording times, or simply be fanatical about how accurate the time-keeper in your PC really is.

The first IBM PC had no independent, self-running real-time clock circuit – each time the PC went through boot-up you had to provide the date and time manually. Time-keeping in your current PC (ever since the IBM-PC/AT) comes in two parts – a real-time clock circuit that runs continuously, independent of the PC's power being on or off, powered by a small battery or storage cell, which is where your PC, well, the operating system, gets its first knowledge of time at start-up; and the clock software within your operating system. Both of these clocks interact on a limited basis.

At start-up, the operating system makes a request of the system-board's BIOS or directly to the real-time clock circuit, to learn what time it is. Once the operating system clock is jump-started it runs by itself as a piece of software keeping time separate from the physical clock circuit. If you change the time in your operating system, the clock software will also reset the time in the clock circuit (keeping you from having to go into BIOS setup to change time twice a year or when you move to a different time zone.)

The real-time clock circuit in your PC costs less than $5 and is about as in-accurate as the give-away watch from a fast-food restaurant. Over the period of a year or more this clock can add or subtract minutes from your day – not a large amount of time but using this reference to catch all

of your favorite shows on your PC-based digital video recorder will cost you the start or end of an episode. The time-keeping software in your operating system can only be as on-time as the real-time clock, plus or minus variances in the timing circuit that makes your CPU tick, adding more inaccuracy to your system.

You could try to improve the accuracy of your PC's time by making a manual correction based on a known time reference source – calling your local phone company's time number, or from radio signals. Either would be a very tedious and inaccurate process. Today the Windows operating system provides for better tools to establish and maintain correct time.

Since my 'early radio days' as a shortwave listener I was fascinated by receiving the then National Bureau of Standards (NBS) (now National Institute of Standards and Technology, NIST) time broadcasts on radio station WWV at 10 MHz. It was interesting to see how accurate my first Timex watch could be from day to day – when a minute or two of inaccuracy seemed tolerable at the age of 12–13. Eventually the correct time became available with a long distance phone call to Boulder/Fort Collins Colorado when you couldn't get near a radio.

The introduction of the Internet and multiple mainframe computers communicating with each other across the country and around the globe in a timely fashion necessitated have a means to synchronize them over the network – thus the Network Time Protocol was born.

The Network Time Protocol, NTP, provides for a server and client to communicate – the server being set from one of many acceptably accurate sources – either a local atomic clock ($20,000

Figure 19-1 *The latest cesium-reference clock used by the National Institute of Standards and Technology – NIST-FI.*

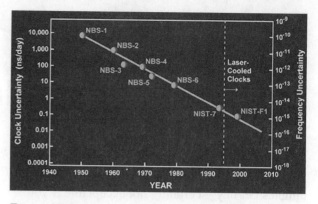

Figure 19-2 *A chart of the accuracy of old to new Cesium-based clocks used at the NIST.*

each), or in the case of the NIST Internet services, the monster master reference clock of all (for the U.S.) dubbed NIST-F1, Figure 19-1 – the seventh such cesium-based atomic reference clock. Client computers would connect to and get time data from an NTP server and the client software adjusts the local computer's date and time settings for both the operating system and the real-time clock circuit.

NIST-F1 drifts very, very, very little. In 2005 the uncertainty of the current NIST reference clock, NIST-F1 was about 5×10^{-16}, which means it would neither gain nor lose a single second in more than 60 million years! Now that's an accurate timepiece!

Before Windows NT, 2000 and now XP there was no accurate way to set the real-time clock in a PC without third-party software. Today every Microsoft operating system has the ability to connect to an NTP server on the Internet or local network with the built-in Windows Time Service. By default Microsoft sets this program to get the time from the time.microsoft.com server.

Configuring the Windows Time Service in Windows XP Home

1. Double-click the clock in the system notification area.

2. Click the "Internet Time" tab in the Date & Time Properties dialog box, Figure 19-3. (If this tab is not visible in the dialog box it means that the Windows Time Service is not running and will have to be re-enabled in the Services control panel.)

3. Select "Automatically synchronize with an Internet time server," if it is not already marked.

4. Choose a server from the drop-down list, or type one that you know yourself from the list below.

5. Click Update Now to test the setting.

If your network time correction worked, your clock will synchronize once a week automatically when connected to the Internet. If not, check your firewall settings to ensure it will allow NTP UDP data traffic on port 123, or try connecting to a different server. As a bit of a time 'nut' I generally prefer to go to an alternate 'source of time' than

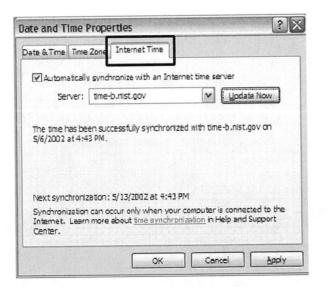

Figure 19-3 *The Internet Time dialog used in Windows XP Home Edition to set the PC clock from an accurate time reference server.*

the time.microsoft.com NTP server, and use the NIST servers themselves, of which there are at least five, or one of the other NTP servers around the U.S.A., listed below:

- time-a.nist.gov – NIST, Gaithersburg, Maryland – 129.6.15.28

- time-b.nist.gov – NIST, Gaithersburg, Maryland – 129.6.15.29

- time-a.timefreq.bldrdoc.gov – NIST, Boulder, Colorado – 132.163.4.101

- time-b.timefreq.bldrdoc.gov – NIST, Boulder, Colorado – 132.163.4.102

- time-c.timefreq.bldrdoc.gov – NIST, Boulder, Colorado – 132.163.4.103

- utcnist.colorado.edu – University of Colorado, Boulder – 128.138.140.44

- time.nist.gov – NCAR, Boulder, Colorado – 192.43.244.18

- time-nw.nist.gov – Microsoft, Redmond, Washington – 131.107.1.10

- nist1.datum.com – Datum, San Jose, California – 66.243.43.21

- nist1.dc.glassey.com – Abovenet, Virginia – 216.200.93.8

- nist1.ny.glassey.com – Abovenet, New York City – 208.184.49.9

- nist1.sj.glassey.com – Abovenet, San Jose, California – 207.126.103.204

- nist1.aol-ca.truetime.com – TrueTime, AOL, Sunnyvale, CA – 207.200.81.113

- nist1.aol-va.truetime.com – TrueTime, AOL, Virginia – 205.188.185.33

Enabling or Disabling the Windows Time Service in Windows XP

If you want to use the built-in NTP time setting service in Windows XP, the Windows Time Service must be enabled with a startup type of Automatic and the service started. If you wish your PC to get time with another application or act as your own NTP time server, you have to reconfigure the Windows Time Service so it will not run as usual. To change the Windows Time Service settings, follow these steps:

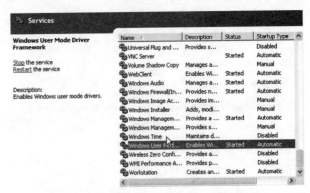

Figure 19-4 *The Windows Services control panel used to enable or disable the built-in Windows Time services.*

1. Right-click My Computer, then select Manage.

2. Double-click Services and Applications, then click Services, Figure 19-4.

3. Scroll down the services list in the right-pane to find then double-click Windows

Time Service, Figure 19-5. At the Startup type: prompt select Automatic (to use the service), or Disabled (to use other time service software) then under Service status: click the Start (to use) or Stop (not use) button. Click OK to close the dialog then close the Services control panel.

Improve PC Time Accuracy With Your Own Time Server

While you could have each PC on your local network 'talk' to the default time.microsoft.com time server to get weekly time corrections, that's not really net-efficient or often enough to keep separate PCs truly in synch and accurate. As your home or business network expands you start to think about consolidating things, reducing your Internet traffic, and tightening up the time accuracy between workstations and servers.

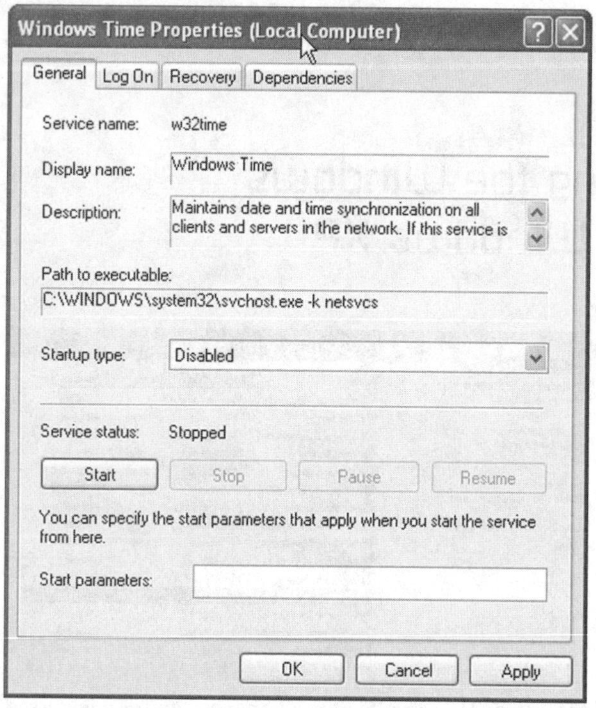

Figure 19-5 *Setting the properties of the Windows Time service to Disabled.*

Time accuracy and synchronization become very important when your computing systems handle large databases, financial transactions and generally anything that must match up or be traceable between points A and B.

The best way to handle time synchronization for your networked systems is to have one system fetch accurate time from a reference server like the NIST, then share that point of calibration with all the other systems on your LAN. A dozen or more different software products can help you do this, but a favorite of mine for years has been Tardis from http://www.kaska.demon.co.uk/ The main Tardis program can fetch and share time data, and comes with a matching NTP client called 'K9' that acts like the Windows Time Service, working in the background to pick up NTP time broadcasts across your local network.

I run Tardis as both NTP client and server on my general purpose 'stuff' server and let it feed time to all of the clients on my network. This keeps Windows Event Logs and other transactions on the same time, making it easier to troubleshoot problems between PCs and servers by reconciling times posted in error logs. I run the companion program to Tardis, K9, on my client PCs to listen for NTP time messages from my Tardis server.

In order for Tardis or any other add-in time software to work correctly, you must disable

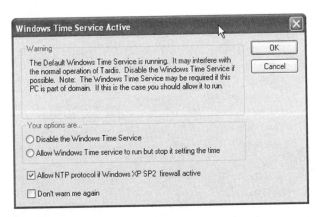

Figure 19-6   *The installation of Tardis detects windows Time Service and prompts you to disable it.*

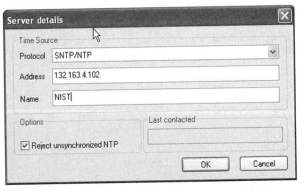

Figure 19-8   *Adding your own server select to Tardis is very easy.*

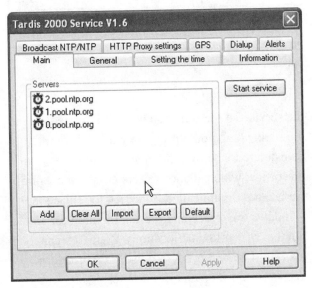

Figure 19-7   *Tardis is all set with three NTP servers to provide the correct time to your PC.*

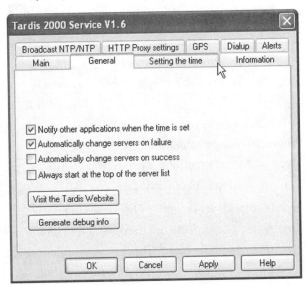

Figure 19-9   *Configuring the behavior of the Tardis time server program.*

Windows Time Service, as indicated above, or let the Tardis installation process disable Windows Time Service for you, Figure 19-6.

'Out of the box' Tardis is configured with three NTP servers, Figure 19-7, run by NTP.ORG, the technical organization that deals with time data and protocols for computing environments. A right click on any of the listed servers allows you to move them up and down the list so you can always start with a preferred server. By clicking the Add button you can configure the program, Figure 19-8, to use different servers to suit your liking.

The General settings tab, Figure 19-9, provides options for how the program operates, including

what to do when the time is set or if obtaining time from one of the servers fails. I would recommend allowing notification of applications when the time is set, changing servers on failure of connection, and always starting at the top of the external server list so you can use a preferred server.

The Setting the time tab, Figure 19-10, provides what I believe are the most significant configuration items to establishing a solid time reference for your network – adjusting the operating system's clock programming to adjust its drift and accuracy, how often to obtain reference time, and automatic adjustments of time gathering and clock frequency.

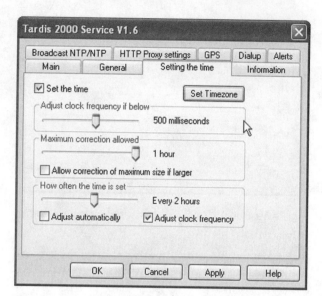

Figure 19-10 *Tardis provides multiple adjustments that affect the accuracy and drift of your PC's time-keeping.*

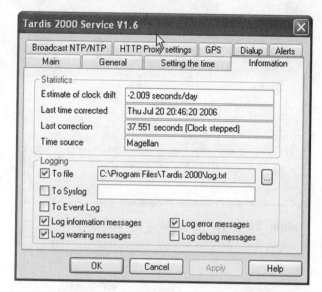

Figure 19-11 *Tardis provides useful diagnostic information about your PC's time accuracy and the correction made.*

When Tardis is running it can tell you, Figure 19-11, the amount of drift in your PC clock and the corrections it makes.

By default, the Tardis application does not broadcast NTP messages. To become your network time server you need to tell Tardis to broadcast NTP data, and how often. The Tardis installation acting as your local should also be set to be an NTP Stratum 1 source.

Whenever you change settings in Tardis you return to the Main tab, Figure 19-7, Stop then Start

the service so the settings take effect immediately. Tardis is not the only program you may want to investigate for setting the time on a single PC. In our office, which supports users around the world, we use a program called ZoneTick to replace the Windows tray clock and present the time for any number of worldwide locations. ZoneTick cannot act as a time server but will use NTP broadcasts to keep your desktop clocks on-time.

Fetching Time From A GPS Receiver

A nice feature of Tardis is that it can take its time reference from a GPS receiver instead of an Internet time server. Using a GPS source for timing data has a lot of advantages if you need to maintain time synchronization even if your Internet connection is lost for a period of time.

To select a GPS instead of an NTP server as the source of your time reference, go to the Main tab, Figure 19-7, select Add. In the Server Details dialog, Figure 19-13, open the pull-down list for Protocol and select the type of GPS source you will be using – NMEA GPS device connected to serial

port being the most common, then you must select the serial/COM port your GPS is connected to.

With a GPS configured as the time source, you must tell Tardis, Figure 19-14, what speed your GPS communicates and add any special commands that 'wake up' your GPS to send data.

For Tardis to use GPS as an accurate time reference your GPS receiver should have clear visibility to more than one GPS satellite for a full 24 hour period. Intermittent or failed GPS reception will decrease the time setting accuracy.

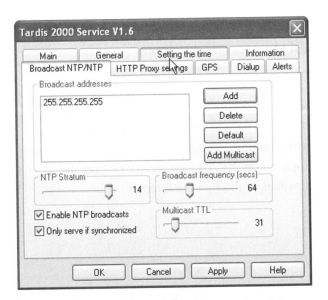

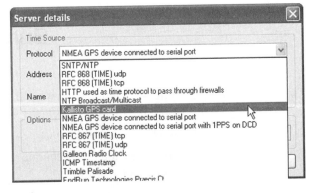

Figure 19-13 *Selecting a GPS receiver as the source of reference time for Tardis.*

Figure 19-12 *Configuring Tardis to be a local time server.*

A Brief History of Time (References)

The National Institute of Standards and Technology web-site offers a tremendous amount of history and information about time measurement, from designs and illustrations of the basic cesium reference clock to the latest mercury atom clock accurate to one second in 70 million years.

Radio buffs of course know about the WWV time broadcasts with highlights of the transmitters at http://tf.nist.gov/timefreq/stations/wwvhistory.htm If you happen to have one of those self-setting 'atomic' alarm clocks or watches (which aren't really atomic) you can see how they get their signals at http://tf.nist.gov/timefreq/stations/wwvb.htm

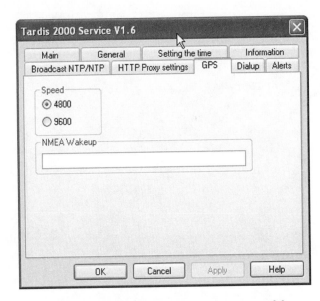

Figure 19-14 *Configuring the COM port speed for Tardis to accept GPS data.*

Summary

Time is a fascinating thing – sometimes we want it to go faster, and sometimes we want it to go slower or not advance at all. No matter how we want time to affect our lives, it should be accurate within and between our PCs and servers.

As we rely more each day on our PCs to bring us and record entertainment, it is important to have our PCs working in synch with broadcast and streaming content sources which base their programming and advertisement start and stop times on the NIST clock services – over the airwaves or through the Internet.

Chapter 20

Securing Your Computer

Personal firewalls, virus protection, and spyware protection are necessary tools against today's wild Internet, but you have to protect your PC as well. If you think using your cat's name or your kid's first initials makes your PC secure from unwanted users? Think again!

With all the regulations and policies thrust upon us at work to use 16-character passwords that cannot be real words, must include mixed-case letters, a few numbers, three special symbol characters, and have to be changed monthly – we want our personal computing to be reasonably pleasant and not a pain to start using. We also want to protect our web experiences – banking, e-mail, and membership sites – to be and remain secure from onlookers and snoops.

When our computers are in use, sometimes we do not want others to be able to install potentially harmful or unknown software, or copy documents to diskettes, CD-ROMs, or USB memory sticks.

There are all sorts of ways to protect our PCs, network, and online accounts, and peripheral devices from abuse. One of them is right at our fingertips, literally, using biometric-based security devices – a relatively inexpensive and darned near foolproof way to protect passwords and our secure accounts. The other is software that can lock-down one or more of the I/O ports on the PC so others cannot tap into the file systems to take or leave files.

In this chapter I present a couple of projects to help you nail down the security of your PC from the outside in, and the inside out. As laptop computers replace desktop systems, and more of us are mobile – working while traveling, relaxing at a local coffee shop or park – theft of these expensive gadgets is on the rise. If theft of a $2000 computer isn't enough of an insult, the high potential for abuse of all the personal information stored on it can ruin your life for years. You really do need to consider how valuable and vulnerable your PC and information are, and the most appropriate steps to mitigate the risks.

Security At Your Finger Tips

The best way to keep people out of your computer is not to let them in, which means using at the very least the password feature that comes with Windows to control and separate user accounts and individual sets of data. When choosing a password, make it complicated. Think beyond words – your keyboard has a LOT of characters to choose from so you are not limited to just the 26 characters of the alphabet (52 counting upper- and lower-case) and 10 numbers, which would limit your total keyboard options to just 62. Add the 32 punctuation symbols and your possible keyboard options total 94.

The more combinations of those 94 keys you can put together the better your password security will be. Most Internet service providers and companies recommend if not enforce that your password include the following complexities:

- At least 6-characters long.

- Your password cannot be or contain common words (as you would find in the dictionary) or common names – Sally, Steve, Jack, etc.

- No repeating letters or numbers – you cannot use '111111' or 'aaaaaa', or even '1234567890' or 'abcdef', etc.

For enhanced security your ISP or employer may recommend or require the following additional complexities:

- Must contain at least one symbol/punctuation mark.

- May not resemble any combination of the characters in your logon name/ID.

- Must be at least 8-characters long.

- Must contain a mix of upper- and lower-case letters.

- Must be changed every 30, 60, 90, or 120 days.

- Must not be a previously password combination.

- Use a pass-phrase instead of a password – 'Mary had a little lamb' is more secure than 'Marylamb'.

System administrators are fond of substituting a select few punctuation symbols for letters so that 'MyCatHasFleas' could be spelled out as 'MyC@Ha$Flea$' using character replacement to mix things up quite a bit.

You might think that 94 different keys used randomly eight times as a password would be pretty secure odds against a computer break-in – after all that makes for 6,000,000,000,000,000 possible combinations – but that level of security is significantly reduced with the tendency to use names, initials, favorite numbers, birthdates, addresses, which most people commonly use as some or all of their password simply because we probably don't have that many brain cells to devote to remembering passwords as there are possible password combinations.

A few things about us are specifically unique – our voice, face, eyes, finger and toe prints, DNA, and a few other details about our lives that all of our friends and work associates may not know. Among these, our fingerprints are one of the most commonly-used unique identifiers we have – and they are pretty handy and very easy to use – just look to any corner of the pages you are reading and imagine the power of the fingers you find there.

Take a quick look at the latest laptops for business use from IBM, Dell and HP, or along the keyboard and mouse aisle at your local computer store, and you will notice more opportunities to use your fingers for something besides typing in your password – those fingers could *be* your password!

Think of the advantages of using just your fingerprint to access your computer and logon to your e-mail or instant messaging accounts – if you are too exhausted to remember your password, just slide your fingertip across or press it onto a fingerprint reader and you are in. (This may be very handy for anyone dedicated to heavy partying with adult beverages, returning home late at night to read e-mail, or engage in online chats.) For parents, using fingerprints is a great way to foil the kids accessing your PC, or using your account to bypass parental controls to surf the Web without restrictions. At work, using your fingerprint for authentication allows you to clean out that growing stack of password-riddled sticky-notes hiding under your keyboard (come on, you knw that's where you hide them!)

Microsoft, Targus, Fellows, DigitalPersona, and APC make fingerprint readers and bundle security software for the PC marketplace. As I looked at these products, like doing a product review or comparison – I found there to be two distinct levels of security available off-the-shelf – from simply replacing the typing-in of logon IDs and passwords for e-mail and web-sites, to more advance versions that can do away with passwords completely by forcing your PC to respond only to a correct fingerprint for logon.

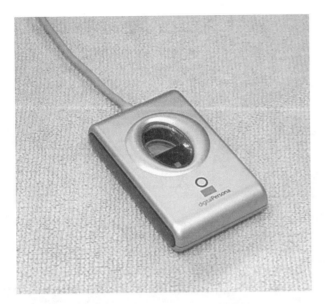

Figure 20-1   *The Digital Persona U.are.U Fingerprint reader.*

On the lightweight side of fingerprint security you'll find the Microsoft Fingerprint Reader using an abbreviated version of the DigitalPersona software, one of the least capable devices as it is useful only for filling out web-site logons and a few other forms once Windows is up and running.

IBM and Dell laptops are available with fingerprint readers and a variety of software features to control PC boot-up, Windows logon and other aspects of system security allowing corporations to maintain very tight security over their PC assets and company information. Couple tight external access security to your PC with encrypting the files on your hard drive, and apart from the possible loss of the data (you do back-up, yes!?) you can feel and be safer.

The APC Biopod with an older, less capable version of Softex's OmniPass software, fills in the middle ground providing for either password or fingerprint access to Windows and the various online sites you set it up for. Softex has a newer, more robust and of course more expensive version of OmniPass intended for commercial applications. The DigitalPersona U.are.U fingerprint reader, Figure 20-1, and software is about as far as you can go in the consumer market to control access to Windows, completely replacing use of your logon and password if you like.

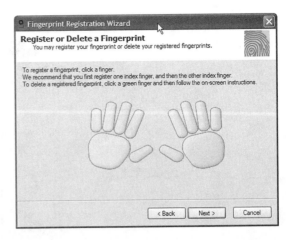

Figure 20-2   *Digital Persona prompts you to register your fingerprints after software installation and restarting Windows.*

As with most USB-connected peripherals, you have to install the software first, then connect the hardware. Because DigitalPersona inserts itself into the Windows logon processes a reboot is required before you begin to use the device.

Upon restart you are given the chance to begin fingerprint registration, Figure 20-2, which will associate your print(s) with your Windows account, and subsequently any web-site or other logons the software can capture.

Select a finger to record, Figure 20-3, then click next to go through the print sensing process, Figure 20-4.

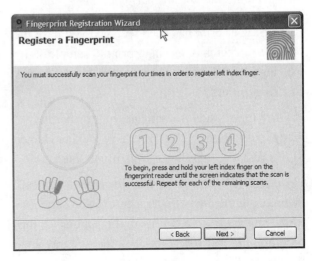

Figure 20-3   *The fingerprint registration prompts you to apply your finger four times to get a good sample.*

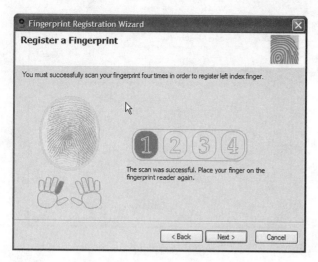

Figure 20-4 *One of four steps DigitalPersona requires to sample a usable fingerprint.*

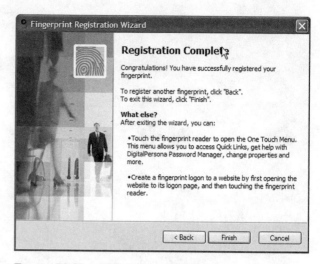

Figure 20-6 *When registration of a print is complete the system is ready to use.*

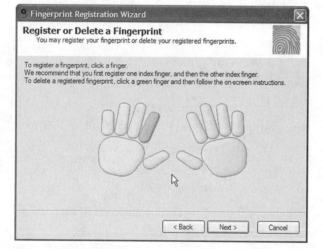

Figure 20-5 *You can record multiple fingerprints to associate with your Windows identification.*

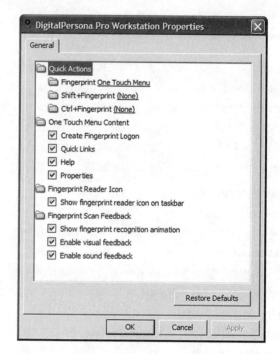

Figure 20-7 *DigitalPersona.*

When your print has been recorded you have the option to select another fingerprint to sample and use, Figure 20-5, should you not be able to use the first fingerprint.

Once registration is complete your reader is ready to use, Figure 20-6. You have another chance to record additional fingerprints if you like.

Once you've registered your fingerprint you can begin using the many fingerprint-enabled services of the application. To configure the options for the program, Figure 20-7, right-click the tool tray icon then select Properties. I find that the visual and sound feedback options are helpful in using the

reader. If you double-click the tool tray icon your computer will lock, as if you pressed the Windows+L keys combination, requiring you to supply a password or fingerprint to unlock the PC.

To establish fingerprint-based logons for website access, simply apply your finger to the reader, then select Create Fingerprint logon, Figure 20-8, and confirm you want to associate a fingerprint with a logon ID for the current web-page, Figure 20-9.

Figure 20-8 *Applying your fingerprint then selecting Create Fingerprint logon begins the process of recording your logon ID and password for web-sites you visit.*

Figure 20-9 *DigitalPersona will confirm the page you've selected to logon to.*

After you associate your logon information with your fingerprint, every visit to the selected web-page will allow you to logon with just a touch of your finger. When DigitalPersona recognizes a web-page it has a fingerprint associated with you get a new icon on your browser's title bar and a prompt, Figure 20-11, letting you know the page is fingerprint-enabled.

In addition to integrating with logons for individual PCs, system administrators can apply DigitalPersona policies from a Windows Active Directory domain controller, with settings that can require a fingerprint only and not allow password logon, as well as creating web-site templates common to all users.

Secure Those Devices

Once your PC is secure, or if you must share a PC and cannot completely secure access to it, you may want to limit what people can do with the PC – specifically whether or not they can copy files to or from CD-ROMs, diskettes, USB memory sticks, etc.

Windows by itself provides no obvious or user-friendly way to lock-out the use of external media devices or ports – USB, CD-ROM, serial, parallel, etc. – once access to Windows is obtained any file available to the user can be copied out of the system and bad files or programs can be transferred in.

A number of new programs provide control over who can use which devices on your PC – from Bluetooth and WiFi wireless devices, to USB and serial ports, to removable media such as CDs and USB FLASH drives.

The simplest to implement is called USBLock from www.advansysperu.com For a mere $22 this program will lock your USB ports, CD-ROM and diskette drives, and secure selected files with a password, Figure 20-12, and provide an ominous warning, Figure 20-13, if any of the secured items has been violated.

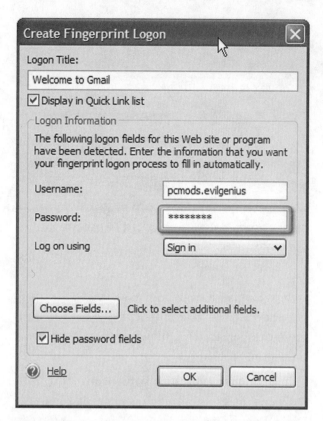

Figure 20-10   *Configuring a web-site logon with DigitalPersona.*

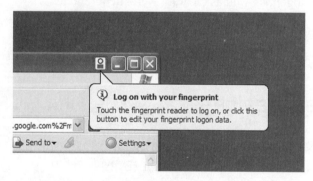

Figure 20-11   *DigitalPersona recognizes your fingeprint-enabled web pages.*

Alternatives to the individual/consumer port security programs are SecureWave's Sanctuary Device Control, and Device Lock form

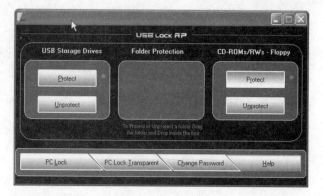

Figure 20-12   *USBLock provides a simple user-interface to secure the most obvious points of I/O device abuse.*

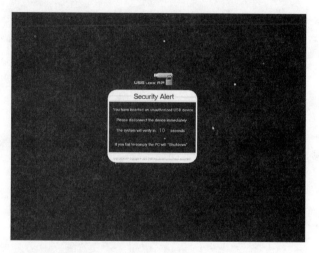

Figure 20-13   *There is no doubt if the security of USBLock is violated as shown by this full-screen intrusion warning.*

www.protect-me.com. These products work with your company's usual network security to provide administrative level control over the ports available to specific users and PCs.

If you are looking for a super-techie brute force way to disable only USB ports, check out Microsoft's Knowledge Base article #823732 which describes ways to control and prevent access to one or more USB devices.

Summary

In this chapter we've explored quite a few ways to secure the average PC from both casual and deliberate intruders who have physical access to your PC. As a matter of course you will also want to be wary of where you leave a laptop computer (in the trunk of your car rather than visible on the front seat), and if you must leave it at your desk secure it with a locking cable available from Targus, Kensington, or similar PC accessory makers.

Protect yourself with a good backup program and media, like the CMS ABS Backup USB drive and backup software, www.cmsproducts.com, or the MirrorFolder automatic data-shadowing utility from www.TechSoftPL.com just in case your PC does slip away from you.

Access Time: The amount of time necessary for data to become available from a disk drive or memory area after a request is issued.

ACPI or Advanced Configuration and Power Interface: A standard specification and method for the monitoring of system activity and control of system configurations with power applied to or removed from system components, or switched to other components, depending on power states. Accommodates different modes of Sleep, Suspend, and Full-On system readiness of many system components.

Adapter: A hardware device, usually a set of connectors and a cable, used between two pieces of equipment to convert one type of plug or socket to another or to convert one type of signal to another. Examples are a 9-25 pin serial port adapter cable, a serial-port–serial-port null modem, and a PC-printer-interface–printer cable.

Adapter Card: A plug-in card used to exchange signals between the computer and internal or external equipment. See also Parallel adapter, Serial adapter, Video adapter, and Disk controller.

Add-in Card: See Adapter card.

Address: A location in memory or on a hardware bus of either a specific piece of data or a physical hardware device.

AGP, Advanced Graphics Port: A high-performance data bus designed specifically to handle digital information from a computer system to a video adapter. AGP is a specific enhancement to the PCI bus, allowing the video adapter to directly access main memory.

Alt-key Codes: A combination of keystrokes using the Alt key plus one or more letter or number keys to cause a particular program function or operation. The Alt key acts like a Shift or Ctrl key to change the function or use of a particular key. Alt-key combinations and their uses differ between many programs. One particular and common use for the Alt key is to allow entry of the decimal value of ASCII characters, especially the upper 128 special characters available with DOS, to draw lines and boxes. These keystrokes require use of any Alt key and the numeric data entry pad (rather than the top-row number keys). One example is pressing and holding the Alt key while entering the number sequence 1, 9, and 7, then releasing the Alt key should cause the entry and display of a set of single crossed-lines the size of a character.

ANSI (American National Standards Institute): A governing body managing specifications for the computer industry and other disciplines. In terms of computing, ANSI maintains a set of standards for the coding and displaying of computer information, including certain 'escape sequences' for screen color and cursor positioning. A device-driver file, [ANSI.SYS], can be loaded in your PC's [CONFIG.SYS] file so that your screen can respond properly to color and character changes provided from programs or terminal sessions between computers.

Antenna: A device used as the terminus for a radio transmitter, converting RF electrical energy into radio waves, and as the 'net' to catch transmitted waves, converting them from wave energy to electrical energy to supply a signal to a radio receiver. Antennas come in various shapes, forms, and sizes with different purposes and effects. An omni-directional antenna radiates and accepts wave energy from all directions equally. A directional antenna radiates and accepts wave energy from only one or a few directions. Antennas, by special tuning and shaping elements, may gain or appear to passively amplify a signal's strength. In a bidirectional amplifier, special circuitry allows signal power to be increased in both the transmitting and receiving directions.

AP, Access Point: A wireless network interface device, acting as/replacing the function of the hub or switch in a wired network, to allow wireless network cards in client systems to connect to a LAN or the Internet.

APM or Advanced Power Management: A standard specification and method for the monitoring of system activity and control of power applied to or removed from system components, accommodating different modes of Sleep, Suspend and Full-On system readiness. Sleep mode allows for the maintaining current system activity with reduced power consumption, such as having disk drives and displays powered off but the CPU and memory retaining the last activities. Suspend mode allows for maintaining minimal if any current system activity with no power consumption.

Application: A computer program or set of programs designed to perform a specific type or set of tasks to make a computer help you to do your work or provide entertainment. Typical applications are games, word processing, database, or spreadsheet programs.

Archive Attribute: See Attributes.

ASCII (American Standard Code for Information Interchange): ASCII defines the numerical or data representation of characters and numbers, and foreign language characters in computer data storage, text files and display. There are 128 predefined characters, numbered 0–127, representing the alphabet, numbers, and data-terminal control functions that nearly any computer system will interpret properly. ASCII characters are represented or transferred in decimal or hexadecimal numeric representations, from 0–255 (decimal) or 00–FFh (hex). The upper 128 characters (128–255) vary between computer systems and languages and are known as the symbol set. IBM defined these as Extended ASCII characters, which include a variety of lines and boxes for pseudo-graphical screen displays. ASCII also defines the format of text files. ASCII text files generated on PCs differ slightly from the original ASCII standard and may appear with extra lines on other computer systems.

Association: The process of wireless adapters establishing a connection with each other on the same radio channel, but not necessarily being able to communicate via TCP/IP or other selected network protocol (this requires authentication). Reassociation occurs if the chosen channel gets too noisy or the signal drops out and picks up again.

ATA: AT-Attachments. An industry-wide specification for the interfacing of devices, typically hard disk drives, to the PC/AT standard data bus.

Attributes: Every DOS file entry, including subdirectories, is accompanied by an attribute byte of information that specifies whether the file is read-only, hidden, system, or archived. Read-only indicates that no program operation should erase or write over a file with this attribute. Hidden indicates that the file should not be displayed or used in normal DOS `DIR`, `COPY` or similar operations. The system attribute indicates that a file belongs to the operating system, which typically applies only to the hidden DOS files `IO.SYS` or `IBMBIO.COM` and `MSDOS.SYS` or `IBMDOS.COM` files. The archive attribute indicates that a file has been changed since the last backup, or that it should be backed-up during the next backup session. Backup operations clear this attribute.

AUTOEXEC.BAT file: An ASCII text file that may contain one or more lines of DOS commands that you want executed every time you boot-up your PC. Also known as just the 'autoexec' file, this file can be customized using a text editor program so that you can specify a DOS prompt, set a drive and directory path to be searched when you call up programs, or load terminate-and-stay resident (TSR) programs that you want to have available all of the time.

Backup: The process of copying one, several, or all of the files on one disk to another disk, a set of diskettes, or tape cartridges for archival storage or routine protection against a system failure or loss of files. A backup should be done regularly and often.

Base Address: The initial or starting address of a device or memory location.

Base Memory: See DOS memory.

Batch File: An ASCII text file that may contain one or more lines of DOS commands that you want to execute by calling for one file, the name of the batch file, rather than keying them in individually. Also known as 'bat' files, these files can be customized using a text editor program so that you can specify a DOS prompt, set a drive and directory path to be searched when you call up programs, or load and execute specific programs. Batch files are used extensively as shortcuts for routine or repetitive tasks or those that you just don't want to have to remember each step for. These files always have the extension .BAT, as required by DOS.

Battery Backup: The facility of retaining power to a system or memory chip from a battery pack when AC power is not available. The battery may be a rechargeable or temporary type.

BIOS (Basic Input/ Output System): The first set of program code to run when a PC system is booted up. The BIOS defines specific addresses and devices and provides software interface services for programs to use the equipment in a PC system. The PC system BIOS resides in a ROM chip on the system board. BIOS also exist on add-in cards to provide additional adapter and interface services between hardware and software.

Bit: A bit is the smallest unit of information or memory possible in a digital or computer system. A bit has only two values: 1, or on, and 0, or off. A bit is the unit of measure in a binary (1/0) system. It might be thought of as a binary information term. A bit is one of 8 pieces of information in a byte, one of 16 pieces in a word (16-bit words), or one of 4 pieces in a nibble (half a byte.)

Blue-screen, Blue Screen of Death, BSOD: The screen appearance commonly associated with the crash or sudden failure of the Windows operating system.

BlueTooth: A short-range 2.4 GHz radio technology designed to simplify communications among various devices. It is most often used for non-network/Internet applications such as remote controls, wireless headsets, mice and keyboards, and printers.

Bootup: The process of loading and running the hardware initialization program to allow access to hardware resources by applications.

Break: See Control break.

Bridge: A network device used to interconnect one or more different networks to act as if they were part of the same network. Bridging different private networks is a typical application. Bridging two Internet connections, or an Internet connection fully onto a LAN together is considered a no-no. In wireless networking a bridge may be two wireless networking devices tied back-to-back to interconnect different wireless LANs, or act as a repeater for client systems.

Buffers: A small area of memory used to temporarily store information being transferred between your computer hardware and a disk drive. This is a settable parameter in the CONFIG.SYS file. Common values range from 3–30, as BUFFERS=x.

Built-in Command: A command or service that loads with and is available as part of the DOS command processor program, COMMAND.COM. DIR, COPY, DEL, TYPE and CLS are examples of some internal DOS commands. See also Internal command and your DOS manual.

Burn-in: The process of running diagnostic or repetitive test software on some or all components of and in a PC system for an extended period of time under controlled conditions. This process helps verify functionality and sort out weak or defective units before they are delivered or used under normal working conditions.

Bus: An internal wiring configuration between the CPU and various interface circuits carrying address, data, and timing information required by one or more internal, built-in, add-in or external adapters, and devices.

Byte: The common unit of measure of memory, information, file size, or storage capacity. A byte consists of 8 bits of information. There are typically two bytes to a word (typically 16 bits) of information. 1024 bytes is referred to as a kilobyte or K, and contains 8192 bits of information.

Cache: A reserved storage area used to hold information enroute to other devices, memory, or the CPU. Information that is called for during a

disk-read operation can be read into a cache with additional information 'stockpiled' ahead of time so that it is available for use faster than having to wait for a disk's mechanical and electronic delays. Caching is becoming common between disks and the computer data bus or CPU and between the memory and CPU to speed up a system's operation. Some CPU chips and controller cards include caching as part of their design.

CDMA, Code-Division Multiple Access: A digital cellular phone technology that uses spread-spectrum techniques. Every channel uses the full available spectrum. Individual conversations are encoded.

CDPD, Cellular Digital Packet Data: A technology for transmitting data over cellular phone frequencies. It uses unused cellular channels in the 800- to 900-MHz range. Data transfer rates of 19.2 kilobits per second are possible.

Checksum: An error-checking method used in file reading and writing operations to compare data sent with checksum information to verify correct reception of the information.

Cluster: The smallest unit of measure of disk storage space under PC or MS-DOS. A cluster typically consists of four or more sectors of information storage space and contains 2048 or more bytes of storage capacity. See Sector.

CMOS Clock: A special clock chip that runs continuously, either from the PC system power supply or a small battery, providing date and time information.

CMOS RAM: A special memory chip used to store system configuration information. Rarely found in PC or XT models and usually found in 286 or higher models.

CMOS Setup: The process of selecting and storing configuration (device, memory, date, and time) information about your system for use during bootup. This process may be performed through your PC's BIOS program or an external (disk-based) utility program.

Command: A word used to represent a program or program function that you want your computer to perform. Commands are issued by you, through the keyboard or mouse, to tell the computer what to do.

Command Line: The screen area immediately after a prompt, where you key in commands to the computer or program. This is most commonly the 'DOS command line', as indicated by the DOS prompt (C>, C:\>, or similar.)

Command-line Editing: The process of changing displayed commands before entering or starting the commanded activity.

Communications Program: An application program that is used to simulate a computer data terminal when communicating with a computer at another location by modem or data communication line. Such programs often provide color display features, modem command setups, telephone number dialing directories, and script or batch-file like automatic keystroke and file transfer functions.

CONFIG.SYS: An ASCII text file that may contain one or more lines of special DOS commands that you want executed every time you bootup your PC. Also known as the 'config' file, this file can be customized using a text editor program, so that you can specify one or more items specific to how your system should operate when it boots up. You may specify device drivers (with DEVICE=) such as memory management programs, disk caching, RAM disks; the number of files and buffers you want DOS to use; the location, name and any special parameters for your command processor (usually COMMAND.COM), among other parameters. Refer to your DOS

manual or device driver software manual for specific information.

Control-Alt-Delete or Ctrl-Alt-Del: The special key sequence used to cause a reboot of a PC system. If there has been no alteration of the special reboot byte code in low memory since the system was turned on, a warm boot or faster reset of the computer will occur. If the reboot code has been changed the system may restart with a complete POST test including RAM memory count. Some systems contain special test code that may be activated in place of POST by setting of the reboot byte and addition of a test jumper on the system board. This latter feature is not well documented and may not be available on all systems.

Control-Break: A combination entry of the Control (Ctrl) and Break (also Pause) keys that can interrupt and stop a program's operation and return the computer to the operating system (DOS). This is also a more robust or stronger version of Ctrl-C key sequence to abort a program. Checking for Control-Break is enhanced by setting BREAK ON in CONFIG.SYS or in DOS. Many programs intercept and do not allow Control-Break to pass to DOS because doing so might cause data loss or corrupt a number of open files in use by a program.

Control-C: A keystroke combination of the Control (Ctrl) and C keys that can interrupt and stop the operation of many programs.

Control Code: A combination of keystrokes used by many programs, or during online sessions to cause special functions or operations to occur. Commonly used control codes are Ctrl-S to stop a display from scrolling so it can be viewed more easily, and Ctrl-Q to cause the display to continue. These commands are entered by pressing the Ctrl key first, then the accompanying single letter code, much like using the Shift or Alt keys to change the action of a letter or number key.

Controller: See Adapter.

Conventional Memory: Also known as DOS memory, this is the range of your PC's memory from 0–640 k where device drivers, DOS parameters, the DOS command processor (COMMAND.COM), your applications programs and data are stored when you use your computer. See Extended, Expanded, Video, High, and Upper Memory.

CPU (Central Processing Unit): The main integrated circuit chip, processor circuit or board in a computer system. For IBM PC-compatible systems the CPU may be an Intel or comparable 8088, 8086, 80286, 80386 (SX or DX), 80486 (SX or DX), Pentium, NEC V20 or V30, or other manufacturer's chip.

Crash: The unexpected and unwanted interruption of normal computer operations. When a program crashes, all open data files may be corrupted or lost, and it is possible that hardware may get 'stuck' in a loop with the computer appearing dead or 'confused' Recovery from a program crash usually requires a reboot or turning off of power for a few seconds, then restarting the system. A disk crash is normally associated with the improper mechanical contact of the read/write heads with the disk platter, although many people consider any disk error or data loss as a crash.

Current Directory: This is the subdirectory you or a program last selected to operate from that is searched first before the DOS PATH is searched when calling a program. See also Current disk drive and Logged drive.

Current Disk Drive: The drive that you have selected for DOS and programs to use before searching the specified drives and directories in the DOS PATH (if any is specified). This may also be the drive indicated by your DOS prompt (typically C>, or C:\>, or similar) or that you have selected by specifying a drive letter followed by a colon

and the Enter key, as in A Enter. This is also known as the logged drive.

Cursor: A line or block character on your system display screen, usually blinking, that indicates where characters that you type will be positioned or where the current prompting for input is active. When at the DOS command line, the cursor is normally at the end of the DOS prompt string.

Decibel, db: A unit of electrical signal measurement using a logarithmic scale used as a reference to quantify radio and audio signals – either power gain, loss, or signal strength. Decibels are measured with special equipment such as spectrum analyzers or may be calculated based on known electrical factors. Typically based on a specific power level (in watts) into a known load impedance (in ohms, 600 ohms for audio, 50 ohms for radio). A reference of 0 decibels is typically 1 mW into a 600 ohm load for audio and 1 mW into a 50 ohm load for radio. Based on a logarithmic scale a –10 db signal is 1/10th and a –3 db signal is 1/2 as strong as a 0 db signal, a 10 db signal is 10 times and a 3 db signal is twice as strong as a 0 db signal. Most radios require between –60 and –90 db to receive a signal clearly. The ambient RF noise in a typical clear, clean reception area ranges from –120 to –100 db, so a receivable signal must be 30-40 db stronger than the noise.

Default: A predetermined or normal value or parameter used by a program or the computer as the selected value, if you do not or can not change it by a command or responding to a prompt for input.

Defragment: The process of reorganizing disk files so that they occupy contiguous sectors and clusters on a disk. This is done to reduce the access time (movement of the data read/write heads) needed to read a single data file.

Destructive Testing: Testing of memory or disk drives that overwrites the original or existing data without regard for restoring it upon completion of the test process.

Device: An actual piece of hardware interfaced to the computer to provide input or accept output. Typical devices are printers, modems, mice, keyboards, displays, and disk drives. There are also some special or virtual devices, handled in software that act like hardware. The most common of these is called NUL, which is essentially nowhere. You can send screen or other output to the NUL device so that it does not appear. The NUL device is commonly used if the actual device to send something to does not exist, but a program requires that output be sent someplace. NUL is a valid 'place' to send output to, although the output really doesn't go anywhere.

Device Driver: A special piece of software required by some hardware or software configurations to interface your computer to a hardware device. Common device drivers are ANSI.SYS, used for display screen control; RAMDRIVE.SYS, which creates and maintains a portion of memory that acts like a disk drive; and HIMEM.SYS, a special device driver used to manage a specific area of extended memory called the high memory area (HMA). Device drivers are usually intended to be used in the CONFIG.SYS file, preceded by a DEVICE= statement. With Windows NT, 2000, Me and XP device drivers are loaded within the operating system structure, sometimes automatically and dynamically.

Diagnostics: Software programs to test the functions of system components.

DIMM (Dual In-line Memory Module): A high-density memory packaging system consisting of 168 pins similar to the edge connector used on larger printed circuit cards. DIMM is used in addition to or in place of SIMM memory design.

DIN Connector: A circular multiwire electronic connector based on international (German) standards. Available in normal and miniature sizes with 3–7 connection pins. The PC uses 5-pin normal and 6-pin-mini DIN connectors for keyboards, and 6-pin mini-DIN connectors for pointing devices.

DIP (Dual In-line Package): A form of integrated circuit housing and connection with two rows of pins on either side of a component body.

DIP Switch: A small board-mounted switch assembly resembling a DIP integrated circuit (IC) package in size and form. Used for the selection of system addresses and options.

Directory: File space on disks used to store information about files organized and referred to through a directory name. Each disk has at least one directory, called the root directory, which is a specific area reserved for other file and directory entries. A hard disk root directory may contain up to 512 other files or directory references, limited by the amount of disk space reserved for root directory entries. The files and directories referred to by the root directory may be of any size up to the limit of available disk space. Directories may be thought of as folders or boxes, as they may appear with some graphical user interfaces, although they are not visually represented that way by DOS. See Root directory and Subdirectories. All directories, except for the root directory, must have a name. The name for a directory follows the 1–8 character restrictions that apply to filenames for DOS-only systems. Windows 95 and higher systems enjoy both longer file and directory names. See also Filename. The term directory has been displaced by folder, though the concept and implementation are the same.

Disk: A rotating magnetic medium used for storing computer files. See also Diskette and Hard Disk.

Disk-Bound Servo Track: The data used by a disk drive to position and verify the location of the data read/write heads. This data may be mixed with the user's data, or on separate data tracks on the disk medium.

Disk Cache: A portion of memory set aside to store information that has been read from a disk drive. The disk cache memory area is reserved and controlled by a disk caching program that you load in `CONFIG.SYS` or `AUTOEXEC.BAT`. The caching program intercepts a program or DOS request for information from a disk drive, reads the requested data, plus extra data areas, so that it is available in memory, which is faster than a disk drive. This is commonly referred to as read-ahead caching. The cache may also be used for holding information to be written to disk, accepting the information faster than the disk can accept it and then writing the information to disk at a short time later.

Disk Drive Adapter: A built-in or add-in card interface or controller circuit that provides necessary connections between the computer system I/O circuits and a disk drive.

Diskette: Also called a floppy diskette, this is a disk medium contained in a cover jacket that can be removed from a disk drive. The term 'floppy' is deemed synonymous or descriptive of the flexible medium that is the magnetically coated disk of thin plastic material.

Disk Label: 1. A surface or sticker on the outside jacket of a diskette that is used for recording information about the contents of the disk. This label may contain as much information as you can write or type in the space provided.

2. A specific area on a disk used to record data as the disk's name or volume label. This area is written with the DOS `LABEL` command, or prompted for input during certain disk format processes. A volume label may be up to 11 characters long. The volume label will appear on-screen during disk directory operations.

References to the disk label may not be clear about which 'label' is to be used. You may use the two definitions above to help determine which label is being referred to by the limitations for each, and the reference you are given.

DLL (Dynamically Linked Library): A file containing executable program functions that are invoked from another program. DLLs may be shared among many applications and are used only when a program requires the functions contained within, reducing program memory and disk space requirements by eliminating duplication of program elements and file size.

DMA (Direct Memory Access): A method of transferring information between a computer's memory and another device, such as a disk drive, without requiring CPU intervention.

DOS (Disk Operating System): A set of software written for a specific type of computer system, disk, file and application types to provide control over disk storage services and other input and output functions required by application programs and system maintenance. All computers using disk drives have some form of disk operating system containing applicable programs and services. For IBM-PC-compatible computers, the term DOS is commonly accepted to mean the computer software services specific to PC systems.

DOS Diskette: A diskette formatted for use with DOS-based PCs and file system.

DOS Memory: Temporary memory used for storage of DOS boot and operating system information, programs, and data during the operation of your computer system. DOS memory occupies up to the first 640 k of Random Access Memory (RAM) space provided in your system's hardware. This memory empties out or loses its contents when your computer is shut off.

DOS System Diskette: A diskette formatted for use with DOS-based PCs and file system that also contains the two DOS-system hidden files and `COMMAND.COM` to allow booting up your system from a diskette drive.

Download: The process of receiving or transferring information from another computer, usually connected via modem, onto your computer system. Downloading is a common method of obtaining public-domain and shareware programs from BBS and online services, obtaining software assistance and upgrades from many companies, or retrieving files or electronic mail from others.

DRAM (Dynamic Random Access Memory): Relatively slow (50 to 200 nsec access time) economical memory integrated circuits. These require a periodic refresh cycle to maintain their contents. Typically used for the main memory in the PC system, but occasionally also used for video memory. See also RAM and SRAM.

Drive: The mechanical and electronic assembly that holds disk storage media and provides the reading and writing functions for data storage and retrieval.

DRM, Digital Rights Management: A generic term referring to various methods of copy or distribution protection for digital information – MP3 music files, CD and DVD-based content and electronically transmitted information. Used to protect copyright and intellectual property distribution to entitle only those authorized to use the information.

DSL (also xDSL, IDSL, ADSL, HDSL): Digital Subscriber Line. A technique of providing high-speed digital communications over conventional telephone wires, using signaling above and different from voice-range frequencies. Implemented in various combinations of upward and downward bandwidth, telephone line and equipment types. Typically lower cost and higher performance than ISDN depending on the implementation. It is possible to carry DSL signaling over some ISDN and Frame Relay circuits for 144–192 kilobit-per-second transfer rates, or on specially conditioned wire pairs to achieve T-1 (1.54 megabit-per-second) data rates.

A Symmetric DSL line can operate as fast as a T-1 line, but the data rate is not guaranteed.

DVD, Digital Versatile Disc: A CD-ROM storage media capable of handling 4.7–17 gigabytes of information. DVD supports rich multi-media information and menu systems to replicate track, scene, and other specific controls to access stored information.

EDGE (Enhanced Data Rates for GSM Evolution): A digital transmission technology that enhances data throughput to about 500 Kbps on GSM cellular data links.

Edge Connector: An electronic connector that is part of the circuit card, made of circuit foil extended to the edge of the board. A circuit card's edge connector mates with the fingers inside a complimentary female socket.

EIA (Electronics Industries Association): An organization that provides and manages standards for many types of electronics designs and implementations. The RS-232C standard for serial data terminal and computer interconnection is the most commonly known EIA standard in the PC market.

EISA (Extended Industry Standard Architecture): The definition of a PC internal bus structure that maintains compatibility with IBM's original PC, XT, and AT bus designs (known as the ISA, or Industry Standard Architecture) but offering considerably more features and speed between the computer system and adapter cards, including a definition for 32-bit PC systems that do not follow IBM's MCA (MicroChannel Architecture).

EMM (Expanded Memory Manager): The term often given to the software or that refers to expanded memory chips and cards. See also Expanded memory.

EMS (Expanded Memory Specification): The IBM-PC-industry standards for software and memory hardware that makes up expanded memory.

ENTER: The command or line termination key, also known as Return on your keyboard. There are usually two Enter keys on your keyboard. Under some applications programs these two keys may have different functions; the numeric keypad Enter key may be used as an 'enter data' key while the alphanumeric keyboard Enter key may be used as a 'carriage return'.

Environment: An area of memory setup and used by the DOS software to store and retrieve a small amount of information that can be shared or referred to by many programs. Among other information that the DOS environment area could hold are the PATH, current drive, PROMPT, COMSPEC, and any SET variables.

Escape Sequence: A set of commands or parameters sent between devices to control operations, printed text orientation or fonts, screen colors and displays, or begin file transfer operations between systems. Many printers accept escape sequences to change typeface or between portrait and landscape modes. Screen displays and the DOS prompt may be controlled by ANSI escape sequences through the device driver ANSI.SYS. These sequences are started with the transmission or issuance of the ASCII ESC character (appearing similar to <-) or the ASCII control code Ctrl-Left Bracket (^[, decimal 27, 1B hex), and follow with lettered or numbered command definitions. A common sequence is ESC-2-j, possibly appearing as ^[2J on your screen, which is the Clear Screen ANSI escape sequence.

ESDI (Enhanced Small Device Interface): A standards definition for the interconnection of older high-speed disk drives. This standard was an alternative to earlier MFM, coincident applications of SCSI, and recent IDE drive interfaces.

EVDO: Evolution-Data Optimized, abbreviated as **EV-DO** or **1xEV-DO** and often **EVDO**, is a broadband wireless data standard by certain cell-phone providers. Data transfer rates range from 400 Kbps to over 5 Mbps depending on the service provider, revision of service offering and radio signal quality.

Executable File: A program file that may be invoked from the operating system. DLLs and overlay files also contain executable program information, but their functions must be invoked from within another program.

Execute: The action that a computer takes when it is instructed to run a program. A running program is said to 'execute' or 'be executing' when it is being used.

Expanded Memory: This is an additional area of memory created and managed by a device driver program using the Lotus-Intel-Microsoft Expanded Memory Specification, known also as LIMS-EMS. There are three common forms of EMS; that conforming to the LIMS-EMS 3.2 standard for software-only access to this memory, LIMS-EMS 4.0 in software, and LIMS-EMS 4.0 in hardware. With the proper hardware, this memory may exist and be used on all PC systems, from PCs to 486 systems. Expanded memory may be made up of extended memory (memory above 1 MB) on 386 and 486 systems, or it may be simulated in extended memory on 286 systems. LIMS-EMS 3.2, 4.0 (software) and 4.0 (hardware) are commonly used for additional data storage for spreadsheets and databases. Only LIMS-EMS conforming to the 4.0 standard for hardware may be used for multitasking. Expanded memory resides at an upper memory address, occupying one 64 k block between 640 k and 1 MB. The actual amount of memory available depends on your hardware and the amount of memory you can assign to be expanded memory. The 64 k block taken up by expanded memory is only a window or port giving access to the actual amount of EMS available. There may be as little as 64 k or as much as 32 MB of expanded memory.

Extended Memory: This is memory in the address range above 1 MB, available only on 80286 or higher systems. It is commonly used for RAM disks, disk caching, and some applications programs. Using a special driver called HIMEM.SYS or similar services provided with memory management software, the first 64 k of extended memory may be assigned as a high memory area, which some programs and DOS can be loaded into.

External Command: A program or service provided as part of DOS that exists as separate programs on disk rather than built into the COMMAND.COM program that loads when you boot-up your system. These programs have .COM or .EXE extensions. Some of these are FORMAT.COM, DISKCOPY.COM, DEBUG.EXE, LABEL.COM, MORE.COM, and PRINT.COM.

FDISK: A special part of the hard disk formatting process required to assign and establish usable areas of the disk as either bootable, active, data-only for DOS, or as nonDOS for other operating system use. The FDISK process is to be performed between the low-level format and the DOS format of a hard disk before its use.

FIFO or FIFO buffering: First-in, first-out. A small capacity data storage element, memory or register that holds data flowing between a source and a destination. The data flow moves in the order in which it is received and cannot be accessed directly or randomly as with normal memory storage. A FIFO is commonly used in serial communication (COM) ports to retain data while applications software and storage devices catch up to and can store the incoming stream of data.

File: An area of disk space containing a program or data as a single unit, referred to by the DOS file directory. Its beginning location is recorded in the file directory, with reference to all space occupied by the file recorded in the DOS file allocation table (FAT). Files are pieces of data or software that you work with on your computer. They may be copied, moved, erased, or modified, all of which is tracked by DOS for the directory and FAT.

File Allocation Table: This is DOS' index to the disk clusters that files or FAT and directories occupy. It provides a table or pointer to the next disk cluster a file occupies. There are two copies of the FAT on a disk, for reliability. When files are erased, copied, moved, reorganized, or defragmented, the FAT is updated to reflect the new position of files or the availability of empty disk space. Files may occupy many different cluster locations on disk, and the FAT is the only reference to where all of the file pieces are.

File Attributes: See Attributes.

Filename: The string of characters assigned to a disk file to identify it. A filename must be at least one, and may be up to eight leading characters as the proper name for DOS-only systems, in which a filename may be followed by a three character extension, separated from the proper name by a period (.). Windows 95, Windows 98 and Windows NT systems may have 'long filenames' of up to 256 characters, including multiple period or 'dot' separators. Allowable filename and extension characters are: A–Z, 0–9, !, @, #, $, ^, &, _, -, {,}, (,), ., ', `, or ~. Also, many of the IBM extended character set may be used. Reserved characters that cannot be used are: %, *, +, =, ;, :, [,], <, >, ?, /, \, |, " and spaces. Filenames must be unique for each file in a directory, but the same name may exist in separate directories. Filenames are assigned to all programs and data files.

Filename Extension: A string of 1–3 characters used after a filename and a separating period (.), with the same character limitations as the filename, for DOS systems. The extension is often used to identify and associate certain types of files to certain applications. DOS uses BAT, EXE, and COM as files it can load and execute, though this does not preclude the use of these extensions for non-executable files. The extensions SYS, DRV and DVR are commonly used for device driver programs that are loaded and used in the CONFIG.SYS file before loading DOS (as COMMAND.COM). Refer to your software documentation for any limitations or preferences it has for filename extensions.

Filespec: Also known as the file specification or file specifier, this is a combination of a drive designation, directory path and filename used to identify a specific file in its exact location on your system's disk drive. References to filespec may appear in examples or as prompts as: d:\path\filename.ext, where d: indicates that you are supposed to place you disk drive information here, \path\ indicates that you should specify the proper directory and subdirectory information here, and filename.ext indicates that you should specify the file's exact name and extension. In use, this might actually be C:\DOS\COM\ FORMAT.COM.

FireWire: Texas Instrument's name-brand for the IEEE-1394 high-speed serial interconnection standard. FireWire connections are typically used between high-end digital video cameras and peripheral storage devices. Also known as iLink (Sony).

Firmware: Software embedded into a device such as a disk drive, video or network adapter, wireless access point or PC card, that controls and supports the functions of the device. The PC's BIOS and the startup code for most computers is firmware specific to the hosting computer board. Firmware resides in either read-only memory chips or in FLASH ROM re-writeable memory chips. The operating system used in PDAs may also be considered firmware.

Fixed Disk: See Hard disk.

Flag: A hardware bit or register, or a single data element in memory that is used to contain the status of an operation, much like the flag on a mailbox signals the mail delivery person that you have an item to be picked up.

Floppy Disk: A slang term. See Diskette.

Format: The process of preparing a disk (floppy or hard) with a specific directory and file structure for use by DOS and applications programs. Formatting may consist of making the disk usable for data storage only, providing reserved space to make the disk bootable later on, or making the disk bootable, including the copying of the DOS hidden files and COMMAND.COM. FORMAT is the final process of preparing a hard disk, preceded by a low-level format and FDISK. All disk media require a format. RAM or virtual disks do not require formatting. Formatting, unless performed with certain types of software, erases all data from a disk.

Fragmentation Threshold: A parameter available in some access point and client wireless devices. If you experience a high packet error rate a slight increase in this value to the maximum of 2432 may help. Too low a value may result in very poor performance.

Frame Relay: A data communications circuit between two fixed points, a user and a Frame Relay routing service, capable of transfer rates between 64 kilobits-per-second up to T-1 rates. May be carried over part of a 'Fractional T-1' circuit.

Gateway: (1) the IP address of the router, switch, cable or DSL modem that your PCs gain access to the Internet or foreign (nonlocal) networks through, (2) network equipment that either bridges, 'repeats', or otherwise relays network traffic from one connection to another.

Gigabyte or GB: A unit of measure referring to 1,024 MB or 1,073,741,824 bytes of information, storage space or memory. Devices with this capacity are usually large disk drives and tape backup units with 1.2 to well over 12 GB of storage area.

GPRS (General Packet Radio Service): A mobile data service available to users of GSM-based cellular phones.

GSM, Global System for Mobile Communications: One of the leading digital cellular phone systems, using narrowband TDMA, which allows eight simultaneous calls on the same radio frequency.

Hard Disk: A sealed disk drive unit with platters mounted inside on a fixed spindle assembly. The actual platter is a hard aluminum or glass surface coated with magnetic storage media. This definition also suits removable hard disks in which the hard platters are encased in a sealed casing and mate with a spindle similar to the attachment of a floppy diskette to the drive motor. The platters are sealed to keep foreign particles from interfering with and potentially damaging the platters or the read/write heads that normally maintain a small gap between them during operation.

Hardware Interrupt: A signal from a hardware device connected to a PC system that causes the CPU and computer program to act on an event that requires software manipulation, such as controlling mouse movements, accepting keyboard input, or transferring a data file through a serial I/O port.

Head Crash: The undesired, uncontrolled mechanical contact of a disk drive's read/write heads with the disk surface. A minor crash may be recoverable with minimal data loss. A severe crash can render a disk or the head assembly completely useless. Minor to severe head crashes may be caused by mechanical shock, excessive vibration, or mishandling of a drive while it is operating. Not all disk errors or loss of data are the result of a physical crash and disk surface damage. Actual head crashes with disk damage are very rare compared with loss of data due to the weakening of magnetic properties of an area of the disk, and program or operational errors.

Hexadecimal: A base-16 numbering system made up of 4 digits or bits of information, where the least significant place equals one and the most significant place equals eight. A hexadecimal, or hex, number is represented as the numbers 0–9 and letters A–F, for the numerical range 0–15 as 0–F. A byte of hex information can represent from 0 to 255 different items, as 00 to FF.

Hidden File: See Attributes.

High Memory Area or HMA: A 64 k region of memory above the 1 MB address range created by HIMEM.SYS or a similar memory utility. The HMA can be used by one program for program storage, leaving more space available in the DOS or the low memory area from 0 to 640 K.

Host Adapter: A built-in or add-in card interface between a device, such as a SCSI hard disk or CD-ROM drive, and the I/O bus of a computer system. A host adapter typically does not provide control functions, instead acting only as an address and signal conversion and routing circuit.

HPFS, High Performance File System: A secure hard disk file system created for OS/2 and extended into the NTFS for Windows NT.

Hub: A network device used to connect several network client devices onto the same network segment. See also Switch.

IBM PC Compatible: A description of a personal computer system that provides the minimum functions and features of the original IBM PC system and is capable of running the same software and using the same hardware devices.

IDC (Insulation Displacement Connector): The type of connector found on flat ribbon cables, used to connect I/O cards and disk drives.

IDE (Integrated Drive Electronics): A standards definition for the interconnection of high-speed disk drives in which the controller and drive circuits are together on the disk drive and interconnect to the PC I/O system through a special adapter card. This standard is an alternative to earlier MFM, ESDI, and SCSI drive interfaces, and it is also part of the ATA-standard.

IEEE-1394: An IEEE-1394 standard for high-speed serial interconnection between computer peripherals – typically cameras and data storage systems. Also known as FireWire™ (Texas Instruments) or iLink™ (Sony).

I/O or Input/Output: The capability or process of software or hardware to accept or transfer data between computer programs or devices.

Interlaced Operation: A method of displaying elements on a display screen in alternating rows of pixels (picture elements) or scans across a display screen, as opposed to non-interlaced operation, which scans each row in succession. Interlacing often indicates a flickering or blinking of the illuminated screen.

Internal Command: A command that loads with and is available as part of the DOS command processor program, COMMAND.COM. DIR, COPY, DEL, TYPE, and CLS are examples of some internal DOS commands. The same as Built-in command. Also see your DOS manual.

Interrupt: See Hardware Interrupt, IRQ, and Software Interrupt.

IRQ (Interrupt Request): This is a set of hardware signals available on the PC add-in card connections that can request prompt attention by the CPU when data must be transferred to/from add-in devices and the CPU or memory.

ISA (Industry Standard Architecture): The term given to the IBM PC, XT, and AT respective 8- and 16-bit PC bus systems. Non–32-bit, non–IBM MicroChannel Architecture systems are generally ISA systems.

ISDN, Integrated Services Digital Network: A technique of providing high-speed digital communications over conventional telephone wires, using signaling above and different from voice-range frequencies. ISDN uses three different signal channels over the same pair of wires, one D-channel for digital signaling such as dialing and several enhanced but seldom used telephone calling features, and 2 B-channels, each capable of handling voice or data communications up to 64 kilobits per second. ISDN lines may be configured as Point-to-Point (both B-channels would connect to the same destination) or Multi-Point (allowing each B-channel to connect to different locations), and Data+Data (B-channels can be used for data-only) or Data+Voice where either B-channel may be used for data or voice transmission. Interconnection to an ISDN line requires a special termination/power unit, known as an NT-1 (network termination 1), which may or may not be built into the ISDN 'modem' or router equipment at the subscriber end. An ISDN 'modem' may be used and controlled quite similarly to a standard analog modem, and may or may not also provide voice-line capabilities for analog devices. An ISDN router must be configured for specific network addresses and traffic control, and may or may not provide voice/analog line capabilities.

ISM, Industrial, Scientific and Medical: ISM applications are the production of physical, biological, or chemical effects such as heating, ionization of gases, mechanical vibrations, hair removal, and acceleration of charged particles. Uses include ultrasonic devices such as jewelry cleaners and ultrasonic humidifiers, microwave ovens, medical devices such as diathermy equipment and magnetic resonance imaging equipment (MRI), and industrial uses such as paint dryers. RF should be contained within the devices but other users must accept interference from these devices. These devices can affect 802.11a and 802.11b services at 2.4 and 5 GHz.

ISO (International Standards Organization): A multifaceted, multinational group that establishes cross-border/cross-technology definitions for many industrial and consumer products. Related to the PC industry, it helps define electronic interconnection standards and tolerances.

Keyboard: A device attached to the computer system that enables manual input of alpha, numeric, and function key information to control the computer or place data into a file.

Kilobyte or K: A unit of measure referring to 1,024 bytes or 8,192 bits of information, storage space, or memory.

Label or Volume Label: A 1–11-character name recorded on a disk to identify it during disk and file operations. The volume label is written to disk with the DOS LABEL or FORMAT programs or with disk utility programs. This may be confused with the paper tag affixed to the outside of a diskette. See Disk label.

LAN (Local Area Network): An interconnection of systems and appropriate software that allows the sharing of programs, data files, and other resources among several users.

Language: The specifically defined words and functions that form a programming language or method to control a computer system. At the lowest accessible level, programmers can control a CPU's operations with assembly language. Applications programs are created initially in different high-level languages such as BASIC, C, or Pascal, which are converted to assembly language for execution. DOS and applications may control the computer's operations with a batch (BAT) processing language or an application-specific macro language.

LCD (Liquid Crystal Display): A type of data display that uses microscopic crystals, which are sensitive to electrical energy to control whether they pass or reflect light. Patterns of crystals may be designed to form characters and figures, as are the small dots of luminescent phosphor in a CRT (display monitor or TV picture tube).

LIMS (Lotus-Intel-Microsoft Standard): See Expanded memory.

Loading High: An expression for the function of placing a device driver or executable program in a high (XMS, above 1 MB) or upper memory area (between 640 K and 1 MB). This operation is performed by a `DEVICEHIGH` or `LOADHIGH` (DOS) statement in the `CONFIG.SYS` or `AUTOEXEC.BAT` file. High memory areas are created by special memory manager programs such as `EMM386` (provided with versions of DOS) and Quarterdeck's `QEMM386`.

Local Bus: A processor to I/O device interface alternative to the PC's standard I/O bus connections, providing extremely fast transfer of data and control signals between a device and the CPU. It is commonly used for video cards and disk drive interfaces to enhance system performance. Local Bus is a trademark of the Video Electronics Standards Association. Local Bus has since been displaced by PCI and AGP.

Logged Drive: The disk drive you are currently displaying or using, commonly identified by the

DOS prompt (`C>` or `A:\>`). If your prompt does not display the current drive, you may do a `DIR` or `DIR/p` to see the drive information displayed.

Logical Devices: A hardware device that is referred to in DOS or applications by a name or abbreviation that represents a hardware address assignment, rather than by its actual physical address. The physical address for a logical device may be different. Logical device assignments are based on rules established by IBM and the ROM BIOS at bootup.

Logical Drive: A portion of a disk drive assigned as a smaller partition of larger physical disk drive. Also a virtual or nondisk drive created and managed through special software. RAM drives (created with `RAMDRIVE.SYS` or `VDISK.SYS`) or compressed disk/file areas (such as those created by older Stacker, DoubleDisk, or SuperStor disk partitioning and management programs) are also logical drives. A 40 MB disk drive partitioned as drives C: and D: is said to have two logical drives. That same disk with one drive area referred to as C: has only one logical drive, coincident with the entire physical drive area. DOS may use up to 26 logical drives. Logical drives may also appear as drives on a network server or mapped by the DOS `ASSIGN` or `SUBST` programs.

Logical Pages: Sections of memory that are accessed by an indirect name or reference rather than by direct location addressing, under control of a memory manager or multitasking control program.

Loopback Plug: A connector specifically wired to return an outgoing signal to an input signal line for the purpose of detecting if the output signal is active or not, as sensed at the input line.

Lower Memory: See DOS memory.

MAC Address: Media Access Control address, a hardware address that uniquely identifies each node of a network. In IEEE 802 networks, the Data Link Control (DLC) layer of the OSI Reference Model is divided into two sublayers: the Logical Link Control (LLC) layer and the Media Access Control (MAC) layer. The MAC layer interfaces directly with the network media. Consequently, each different type of network media requires a different MAC layer.

MAN, Metropolitan Area Network: A network connection between two locations, typically a T-1 circuit but may be ISDN, Frame Relay or other (possibly a VPN over any Internet connection type) used to bridge Local Area Networks in related office facilities. There is typically a shorter distance between locations than a WAN, as within a city or community.

Math Coprocessor: An integrated circuit designed to accompany a computer's main CPU and speed floating point, and complex math functions that would normally take a long time if done with software and the main CPU. Allows the main CPU to perform other work during these math operations.

Megabyte or MB: A unit of measure referring to 1,024 k or 1,048,576 bytes of information, storage space, or memory. One megabyte contains 8,388,608 bits of information. One megabyte is also the memory address limit of a PC- or XT-class computer using an 8088, 8086, V20, or V30 CPU chip. 1 MB is 0.001 GB.

Megahertz or MHz: A measure of frequency in millions of cycles per second. The speed of a computer system's main CPU clock is rated in megahertz.

Memory: Computer information storage area made up of chips (integrated circuits) or other components, which may include disk drives. Personal computers use many types of memory, from dynamic RAM chips for temporary DOS, extended, expanded, and video memory to static

RAM chips for CPU instruction caching to memory cartridges and disk drives for program and data storage.

Memory Disk: See RAM disk.

Microprocessor: A computer central processing unit contained within one integrated circuit chip package.

MIDI (Musical Instrument Device Interface): An industry standard for hardware and software connections, control, and data transfer between like-equipped musical instruments and computer systems.

Milliwatt, mW: A unit of power measurement equal to one-thousandth of a watt. Most unlicensed and 'Part 15' devices (FRS walkie-talkies) have a transmitted power limit of 100 mW. A portable cellular telephone transmitter output is typically 600 mW.

Modem: An interface between a computer bus or serial I/O port and wiring, typically a dial-up telephone line, used to transfer information and operate computers distant from each other. Modem stands for modulator/demodulator. It converts computer data into audible tone sounds that can be transferred by telephone lines to other modems that convert the tone sounds back into data for the receiving computer. Early modems transfer data at speeds of 110 to 300 bits per second (11 to 30 characters per second). Recent technology allows modems to transfer data at speeds of 56,700 bits (5,670 characters or bytes) per second and higher, often compressing the information to achieve these speeds and adding error correction to protect against data loss due to line noise. Modems typically require some form of UAR/T (Universal Asynchronous Receiver/Transmitter) as the interface to the computer bus.

Motherboard: The main component or system board of your computer system. It contains the necessary connectors, components, and interface circuits required for communications between the CPU, memory, and I/O devices.

Multi-path: Multiple reflections of an RF signal between a receiver and transmitter that can often cause multiple signals to arrive at the receiving station at the same time, occasionally canceling out each other and the main, direct line-of-sight signal. Multi-path instances are one of the major causes of failure of wireless networking.

Multitasking: The process of software control over memory and CPU tasks allowing the swapping of programs and data between active memory and CPU use to a paused or nonexecuting mode in a reserved memory area, while another program is placed in active memory and execution mode. The switching of tasks may be assigned different time values for how much of the processor time each program gets or requires. The program you see onscreen is said to be operating in the foreground and typically gets the most CPU time, while any programs you may not see are said to be operating in the background, usually getting less CPU time. DESQview and Windows are two examples of multitasking software in common use on PCs.

NAN, Neighborhood Area Network: Typically an ad hoc wireless network installed by a neighbor with an 802.11x access point at a location providing a high-speed Internet connection (cable, DSL, T-1 or other wireless service) to provide wireless Internet access within a block or two of home. With greater coverage a NAN may also be considered a 'CAN', a Community or Campus Area Network.

Network: The connection of multiple systems together or to a central distribution point for the purpose of information or resource sharing.

Network Interface Card or NIC: Typically an ISA, PCI or PC Card plug-in adapter used to connect a wired network to a computer. Wireless NICs are used to replace the wires.

Nibble: A nibble is one-half of a byte, or 4 bits, of information.

Ni-cad Battery: An energy cell or battery composed of nickel and cadmium chemical compositions, forming a rechargeable, reusable source of power for portable devices.

Noninterlaced Operation: A method of displaying elements on a display screen at a fast rate throughout the entire area of the screen, as opposed to interlaced operation, which scans alternate rows of display elements or pixels, the latter often indicating a flickering or blinking of the illuminated screen.

Norton or Norton Utilities: A popular suite of utility programs used for PC disk and file testing and recovery operations, named after their author, Peter Norton. The first set of advanced utilities available for IBM PC–compatible systems.

NTFS, NT File System: The NT File System for hard disk drives in Windows NT, 2000 and XP workstations and servers provides security and recoverability, using a secure indexed file structure linked to the Security Access Manager of the operating system. It is nonreadable by any version of DOS.

Null Modem: A passive, wire-only data connection between two similar ports of computer systems, connecting the output of one computer to the input of another, and vice versa. Data flow control or handshaking signals may also be connected between systems. A null modem is used between two nearby systems much as you might interconnect two computers at different locations by telephone modem.

Offsets: When addressing data elements or hardware devices, often the locations that data are stored or moved through is in a fixed grouping, beginning at a known or base address, or segment of the memory range. The offset is that distance, location, or number of bits or bytes that the desired information is from the base or segment location. Accessing areas of memory is done with an offset address based on the first location in a segment of memory. For example, an address of 0:0040h represents the first segment, and an offset of 40 bytes. An address of A:0040h would be the 40th (in hex) byte location (offset) in the tenth (Ah) segment.

Omnidirectional Antenna: An antenna that receives and transmits in all directions equally. Some omnidirectional antennas are constructed to concentrate the transmitted and received signals into a narrow horizontal pattern to create passive amplification or gain for the signals.

Online: A term referring to actively using a computer or data from another system through a modem or network connection.

Online Services: These are typically commercial operations much like a BBS that charge for the time and services used while connected. Most online services use large computers designed to handle multiple users and types of operations. These services provide electronic mail, computer and software support conferences, online game playing, and file libraries for uploading and downloading public-domain and shareware programs. Often, familiar communities or groups of users form in conferences, making an online service favorite or familiar places for people to gather. Access to these systems is typically by modem, to either a local data network access number or through a WATS or direct-toll line. America Online, Prodigy, and CompuServe are among the remaining online services available in the United States and much of the world at large. Online services have given way to the World Wide Web and portal sites such as Yahoo! and MSN.

Operating System: See Disk operating system.

OS/2: A 32-bit operating system, multitasking control, and graphical user interface developed by Microsoft, currently sold and supported by IBM. OS/2 allows the simultaneous operation of many DOS, Windows, and OS/2-specific application programs.

OSS, Operational Support Systems: A term originally coined by telephone companies to describe the systems used to provision, manage, and bill for telephone-related services. Today such systems include customer relationship management and workforce administration. In relation to wireless networking, these systems tie together customer orders, installations, customer support, and service maintenance record-keeping.

Overlays: A portion of a complete executable program, existing separately from the main control program, that is loaded into memory only when it is required by the main program, thus reducing overall program memory requirements for most operations. Occasionally, overlays may be built into the main program file, but they are also not loaded into memory until needed. Overlays per se have been made obsolete by Windows and DLLs.

Page Frame: The location in DOS/PC system memory (between 640 k and 1 MB) where the pages or groups of expanded memory are accessed.

Parallel I/O: A method of transferring data between devices or portions of a computer where eight or more bits of information are sent in one cycle or operation. Parallel transfers require eight or more wires to move the information. At speeds from 12,000 to 92,000 bytes per second or faster, this method is faster than the serial transfer of data where one bit of information follows another. Commonly used for the printer port on PCs.

Parallel Port: A computer's parallel I/O (LPT) connection, built into the system board or provided by an add-in card.

Parameter: Information provided when calling or within a program specifying how or when it is to run with which files, disks, paths, or similar attributes.

Parity: A method of calculating the pattern of data transferred as a verification that the data has been transferred or stored correctly. Parity is used in all PC memory structures, as the 9th, 17th, or 33rd bit in 8-, 16-, or 32-bit memory storage operations. If there is an error in memory, it will usually show up as a parity error, halting the computer so that processing does not proceed with bad data. Parity is also used in some serial data connections as an eighth or ninth bit to ensure that each character of data is received correctly.

Partition: A section of a hard disk drive typically defined as a logical drive, which may occupy some or all of the hard-disk capacity. A partition is created by the DOS FDISK or other disk utility software.

Path: A DOS parameter stored as part of the DOS environment space, indicating the order and locations DOS is to use when you request a program to run. A path is also used to specify the disk and directory information for a program or data file. See also Filespec.

PC: The first model designation for IBM's family of personal computers. This model provided 64 to 256 k of RAM on the system board, a cassette tape adapter as an alternative to diskette storage, and five add-in card slots. The term generally refers to all IBM PC–compatible models and has gained popular use as a generic term referring to all forms, makes, and models of personal computers.

PC Compatible: See IBM PC compatible and AT compatible.

PCMCIA: Personal Computer Memory Card Industry Association. An I/O interconnect definition used for memory cards, disk drives, modems, network and other connections to portable computers. The term has been displaced by the use of *PC Card* instead.

PDA, Personal Digital Assistant: Typically a hand-held device used as an electronic address book, calendar and notepad. Commonly using the Palm OS, Windows CE or similar dedicated operating system.

Pentium: A 64-bit Intel microprocessor capable of operating at 60–266+MHz, containing a 16 k instruction cache, floating point processor, and several internal features for extremely fast program operations.

Pentium II: A 64-bit Intel microprocessor capable of operating at 200–450+MHz, containing a 16 k instruction cache, floating point processor, and several internal features for extremely fast program operations. Packaged in what is known as Intel's 'Slot 1' module, containing the CPU and local chipset components.

Pentium III: A 64-bit Intel microprocessor capable of operating at 450–800+MHz. Packaged in what is known as Intel's 'Slot 1' module, containing the CPU and local chipset components.

Peripheral: A hardware device internal or external to a computer that is not necessarily required for basic computer functions. Printers, modems, document scanners, and pointing devices are peripherals to a computer.

Peripheral Component Interconnect or PCI: An Intel-developed standard interface between the CPU and I/O devices providing enhanced system performance. PCI is typically used for video and disk drive interconnections to the CPU.

Physical Drive: The actual disk drive hardware unit, as a specific drive designation (A:, B:, or C:, etc.), or containing multiple logical drives, as with a single hard drive partitioned to have logical drives C:, D:, and so on. Most systems or controllers provide two to four physical floppy diskette drives and up to two physical hard disk drives, which may have several logical drive partitions.

Pixel: Abbreviation for picture element. A single dot or display item controlled by your video adapter and display monitor. Depending on the resolution of your monitor, your display may have the ability to display 320×200, 640×480, 800×600, or more picture elements across and down your monitor's face. The more elements that can be displayed, the sharper the image appears.

Plug and Play: A standard for PC BIOS peripheral and I/O device identification and operating system configuration established to reduce the manual configuration technicalities for adding or changing PC peripheral devices. Plug-and-Play routines in the system BIOS work with and around older, legacy, or otherwise fixed or manually configured I/O devices and reports device configuration information to the operating system. (The operating system does not itself control or affect PnP or I/O device configurations.)

PnP: See Plug and Play.

PoE, Power Over Ethernet: A wiring method to add DC power supply to standard Ethernet cabling to power an Ethernet device, typically a wireless access point or amplifier, without having to add separate power cabling to the interconnection.

Pointing Device: A hardware input device, a mouse, trackball, cursor tablet, or keystrokes used to direct a pointer, cross-hair, or cursor position indicator around the area of a display screen to locate or position graphic or character elements or select position-activated choices (buttons, scroll bar controls, menu selections, etc.) displayed by a computer program.

Port Address: The physical address within the computer's memory range that a hardware device is set to decode and allow access to its services through.

POST (Power On Self Test): A series of hardware tests run on your PC when power is turned on to the system. POST surveys installed memory and equipment, storing and using this information for bootup and subsequent use by DOS and applications programs. POST will provide either speaker beep messages, video display messages, or both if it encounters errors in the system during testing and bootup.

PPP, Point-to-Point Protocol: a method of connecting a computer, typically by serial port connection or modem, to a network. The method used to create a dial-up TCP/IP connection between your computer and your Internet Service Provider.

Program, Programming: A set of instructions provided to a computer specifying the operations the computer is to perform. Programs are created or written in any of several languages that appear at different levels of complexity to the programmer or in terms of the computer itself. Computer processors have internal programming known as microcode that dictates what the computer will do when certain instructions are received. The computer must be addressed at the lowest level of language, known as machine code, or one that is instruction specific to the processor chip being used. Programming is very rarely done at machine-code levels except in development work.

The lowest programming level that is commonly used is assembly language, a slightly more advanced and easier-to-read level of machine code, also known as a second-generation language. Most programs are written in what are called third-generation languages such as BASIC, Pascal, C or FORTRAN, more readable as a text file. Batch files, macros, scripts, and database programs are a form of third-generation programming language specific to the application or operating with which system they are used. All programs are either interpreted by an intermediate application or compiled with a special program to convert the desired tasks into machine code.

Prompt: A visual indication that a program or the computer is ready for input or commands. The native DOS prompt for input is shown as the a disk drive letter and 'right arrow', or 'caret', character (C>). The DOS prompt may be changed with the DOS PROMPT internal command, to indicate the current drive and directory, include a user name, the date or time, or more creatively, flags or colored patterns.

Public Domain: Items, usually software applications in this context, provided and distributed to the public without expectation or requirement of payment for goods or services, although copyrights and trademarks may be applied. Public-domain software may be considered as shareware, but shareware is not always in the public-domain for any and all to use as freely as they wish.

RAM (Random Access Memory): A storage area that information can be sent to and taken from by addressing specific locations in any order at any time. The memory in your PC and even the disk drives are a form of random access memory, although the memory is most commonly referred to as the RAM. RAM memory chips come in two forms, the more common Dynamic RAM (DRAM), which must be refreshed often to retain the information stored in it, and Static RAM, which can retain information without refreshing, saving power and time. RAM memory chips are referred to by their storage capacity and maximum speed of operation in the part numbers assigned to them. Chips with 16 k and 64 k capacity were common in early PCs, 256 k and 1 MB chips in the early 1990s, but 8, 16, 32 and 64 megabyte RAM components are now more common.

RAM Disk or RAM Drive: A portion of memory assigned by a device driver or program to function like a disk drive on a temporary basis. Any data stored in a RAM drive exists there as long as your computer is not rebooted or turned off.

Read Only: An attribute assigned to a disk file to prevent DOS or programs from erasing or writing over a file's disk space. See Attributes.

Refresh: An internal function of the system board and CPU Memory refresh timing circuits to recharge the contents of Dynamic RAM so that contents are retained during operation. The standard PC RAM refresh interval is 15 microseconds. See also DRAM, RAM, SRAM, and Wait states.

RETURN: See Enter key.

ROM (Read-Only Memory): This is a type of memory chip that is preprogrammed with instructions or information specific to the computer type or device it is used in. All PCs have a ROM-based BIOS that holds the initial bootup instructions that are used when your computer is first turned on or when a warm-boot is issued. Some video and disk adapters contain a form of ROM-based program that replaces or assists the PC BIOS or DOS in using a particular adapter.

ROM BIOS: The ROM chip-based start-up or controlling program for a computer system or peripheral device. See also BIOS and ROM.

Root Directory: The first directory area on any disk media. The DOS command processor and any CONFIG.SYS or AUTOEXEC.BAT file must typically reside in the root directory of a bootable disk. The root directory has space for a fixed number of entries, which may be files or subdirectories. A hard disk root directory may contain up to 512 files or subdirectory entries, the size of which is limited only by the capacity of the disk drive. Subdirectories may have nearly unlimited numbers of entries.

Router: A network interface device used to connect and control the path data can take between one or more devices, over one or more connection paths. Typically used at the subscriber end of a T-1, DSL, cable, or other high-speed connection. As an example, an office may have a DSL connection to the Internet and a private/dedicated T-1 to a remote office, and a connection to the office LAN – the router 'decides' or is told to transfer Internet traffic (browsing, etc.) to the DSL circuit only and office LAN traffic to the other office over the T-1 only, but prevent Internet traffic from appearing on the private T-1 while keeping private T-1 traffic off the Internet. Effectively the two office LANs become virtually bridged while the Internet traffic is routed onto the LAN only.

RSSI, Received Signal Strength Indicator: A feature of many wireless integrated circuits to provide a means of measuring the relative strength of the signals you are receiving.

RTS Threshold: A configurable parameter available in some access point and client wireless devices. This parameter controls what size data packet the low level RF protocol issues to a Request to Send (RTS) packet. Default is 2432. Setting this parameter to a lower value causes RTS packets to be sent more often, consuming more of the available bandwidth, reducing the apparent throughput. The more often RTS packets are sent the quicker the system can recover from interference or collisions.

SATA (Serial ATA, Serial AT-Attachment): An interface connection between the host adapter/disk-controller in a computer and disk drives. Rapidly replacing Parallel ATA (standard IDE) as it is capable of doubled data transfer rates.

SCSI (Small Computer System Interface): An interface specification for interconnecting peripheral devices to a computer bus. SCSI allows for attaching multiple high-speed devices such as disk and tape drives through a single cable.

Segments: A method of grouping memory locations, usually in 64 k increments or blocks, to make addressing easier to display and understand. Segment 0 is the first 64 k of RAM in a PC. Accessing areas of memory within that segment is done with an offset address based on the first location in the segment. An address of 0:0040h would be the 40th (in hex) byte location in the first 64 k of memory. An address of A:0040h would be the 40th (in hex) byte location in the tenth (Ah) 64 k of memory.

Serial I/O: A method of transferring data between two devices one bit at a time, usually within a predetermined frame of bits that makes up a character, plus transfer control information (start and stop or beginning and end of character information). Modems and many printers use serial data transfer. One-way serial transfer can be done on as few as two wires, with two-way transfers requiring as few as three wires. Transfer speeds of 110,000 to 115,000 bits (11,000 to 11,500 characters) per second are possible through a PC serial port.

Serial Port: A computer's serial I/O (COM) connection, built into the system board or provided by an add-in card.

Shadow RAM: A special memory configuration that remaps some or all of the information stored in BIOS and adapter ROM chips to faster dedicated RAM chips. This feature is controllable on many PC systems that have it, allowing you to use memory management software to provide this and other features.

Shareware: Computer applications written by noncommercial programmers, offered to users with a try-before-you-buy understanding, usually with a requirement for a registration fee or payment for the service or value provided by the application. This is very much like a cooperative or user-supported development and use environment, as opposed to buying a finished and packaged product off the shelf with little or no opportunity to test

and evaluate if the application suits your needs. Shareware is not public-domain software. Payment is expected or required to maintain proper, legal use of the application.

SIMM (Single Inline Memory Module): A dense memory packaging technique with small memory chips mounted on a small circuit board that clips into a special socket.

SIP (Single Inline Package): Typically a dense memory module with memory chips mounted on a small circuit board with small pins in a single row that plugs into a special socket.

Software Interrupt: A (nonhardware) signal or command from a currently executing program that causes the CPU and computer program to act on an event that requires special attention, such as the completion of a routine operation or the execution of a new function.

Many software interrupt services are predefined and available through the system BIOS and DOS, while others may be made available by device driver software or running programs. Most disk accesses, keyboard operations, and timing services are provided to applications through software interrupt services.

Spectrum Analyzer: A piece of expensive specialized test equipment used to view and measure a variety of signals within a narrow or broad spectrum. Typically used by engineers to help design and tune radio equipment, or survey sites for radio interference.

SRAM (Static Random Access Memory): Fast access (less than 50 nanoseconds), somewhat expensive, memory integrated circuits that do not require a refresh cycle to maintain their contents. Typically used in video and cache applications. See also DRAM and RAM.

ST506/412: The original device interface specification for small hard drives, designed by

Seagate and first commonly used in the IBM PC/XT.

Start Bit: The first data bit in a serial data stream, indicating the beginning of a data element. In the 'old' days when mechanical tele-printers were used the Start bit signaled the motor and mechanical elements of the printer to start running.

Stop Bit: The last data bit or bits in a serial data stream, indicating the end of a data element. Like the Start bit, the Stop bit signaled the time when the teleprinter mechanics should stop running.

Subdirectory: A directory contained within the root directory or in other subdirectories, used to organize programs and files by application or data type, system user, or other criteria. A subdirectory is analogous to a file folder in a filing cabinet or an index tab in a book. The concept is the same but the term subdirectory has been displaced by folder.

Surface Scan: The process of reading and verifying the data stored on a disk to determine its accuracy and reliability, usually as part of a utility or diagnostic program's operation to test or recover data.

Switch: A network device, much like a hub, that interconnects several network devices onto the same network segment, but that can automatically keep interdevice traffic within its own network segment to reduce overall LAN traffic. In effect, switches can act like routers without complex router setup and instructions.

Sysop: The system operator of a BBS, on-line service forum, or network system. Now we call them system or network administrators.

System Attribute or System File: See Attributes.

T-1: A 1.5 megabits-per-second high-speed four-wire data or voice circuit used to convey multiple channels of data or voice traffic between two points. A T-1 line is generally expensive and requires special 'modem' equipment to support interconnection to computers, routers or telephone systems. In voice service a T-1 line carries 23 64 kilobyte-per-second channels of discrete call information and voice traffic.

TDMA, Time Division Multiple Access: A technology for delivering digital wireless service, typically related to digital cellular telephone services. TDMA divides a radio frequency signal into time slots and then allocates slots to multiple calls. With TDMA a single frequency can support multiple, simultaneous data channels. (Author's note – depending on how the cellular system operator set the system and how crowded it is, TDMA systems often sound robotic/synthesized or distorted compared to CDMA cellular systems.)

TFTP, Tiny File Transfer Protocol: A nonsecure data communication protocol similar to the Internet's FTP that is used to transfer firmware or operating parameters to dedicated devices such as routers and firewalls.

TorX Fastener and Tool: A four-point special fastener and tool for same that differs from a normal slotted/flat edge, cross-head, or hexagonal fastener.

TSR (Terminate-and-Stay-Resident Program): Also known as a memory-resident program. A program that remains in memory to provide services automatically or on request through a special key sequence (also known as hot keys). Device drivers (MOUSE, ANSI, SETVER) and disk caches, RAM disks, and print spoolers are forms of automatic TSR programs. SideKick, Lightning, and assorted screen-capture programs are examples of hot-key–controlled TSR programs. Under Windows TSRs or resident programs run at the time and within the same operating system control as other programs.

Twisted-pair Cable: A pair of wires bundled together by twisting or wrapping them around each other in a regular pattern. Twisting the wires reduces the influx of other signals into the wires, preventing interference, as opposed to coaxial (concentric orientation) or parallel wire cabling.

UAR/T (Universal Asynchronous Receiver/Transmitter): This is a special integrated circuit or function used to convert parallel computer bus information into serial transfer information and vice versa. A UAR/T also provides proper system-to-system online status, modem ring and data carrier detect signals, as well as start/stop transfer features. The most recent version of this chip, called the 16550A, is crucial to high-speed (greater than 2400 bits per second) data transfers under multitasking environments.

UNIX: A high performance multi-tasking operating system designed by AT&T/Bell Laboratories in the late 1960s. Today UNIX has several offshoots and derivatives, including LINUX, Sun OS/Solaris, FreeBSD and others. UNIX is the operating system of choice for many 'enterprise' business applications, and most of the servers and Internet services we enjoy today.

Upload: The process of sending or transferring information from your computer to another, usually connected by modem or a network. Uploading is done to BBS and online services when you have a program or other file to contribute to the system or to accompany electronic mail you send to others.

Upper Memory and Upper Memory Blocks: Memory space between 640 k and 1 MB that may be controlled and made available by a special device or UMB (`EMM386.SYS`, `QEMM386`, `386Max`, etc.) for the purpose of storing and running TSR programs and leaving more DOS RAM (from 0 to 640 k) available for other programs and data. Some of this area is occupied by BIOS, video, and disk adapters.

USB 'key' (aka USB 'drive' aka USB 'stick'): A portable data storage device containing a FLASH memory module in sizes from 64 MB to 4 GB, that acts like a removable disk drive. USB keys can be made bootable and contain an operating system useful for running utility programs.

USB, Universal Serial Bus: A high-speed two-wire interconnection between a host system and up to 256 discrete separate devices. USB allows for connection, disconnection, and reconnection of peripheral devices from a computer, supported by Plug and Play or a similar autoconfiguring device support system. USB is commonly used to connect printers, cameras, scanners, and some network devices to PCs and Apple Macintosh computers. It is supported in Windows 98, 98SE, Me, 2000, and XP as well as Mac OS9.x, OS X, and later versions of Linux.

Utilities: Software programs that perform or assist with routine functions such as file backups, disk defragmentation, disk file testing, file and directory sorting, etc. See also Diagnostics.

Variable: Information provided when calling or within a program specifying how or when it is to run with which files, disks, paths, or similar attributes. A variable may be allowed for in a batch file, using %1 through %9 designations to substitute or include values keyed-in at the command line when the Batch file is called.

VGA (Video Graphics Array): A high-resolution text and graphics system supporting color and previous IBM video standards using an analog-interfaced video monitor.

Video Adapter Card: The interface card between the computer's I/O system and the video display device.

Video Memory: Memory contained on the video adapter dedicated to storing information to be processed by the adapter for placement on the display screen. The amount and exact location of video memory depends on the type and features of your video adapter. This memory and the video adapter functions are located in upper memory between 640 k and 832 k.

Virtual Disk: See RAM disk.

Virtual Memory: Disk space allocated and managed by an operating system that is used to augment the available RAM memory, and is designed to contain inactive program code and data when switching between multiple computer tasks.

VLAN, Virtual Local Area Network: An interconnection between devices (client PCs, servers, printers, etc.) as if they were part of another LAN some distance away. Typical of a WLAN connection used for making a VPN connection to a LAN, with the ability to roam between different WLAN connections and still be part of the LAN.

Volume Label: See Disk Label and Label.

VPN, Virtual Private Network: An encrypted connection from a client workstation, a server or Local Area Network, to another server or LAN at a different location over the public Internet. Typically used for telecommuting or roaming workers accessing a corporate network, but also useful in securing Wireless LAN connections. May also be used to create a VLAN.

Wait States: A predetermined amount of time between the addressing of a portion of a memory location and when data may be reliably read from or written to that location. This function is controlled by the BIOS, and it is either permanently set or changed in CMOS setup. Setting this parameter too low may cause excessive delays or unreliable operation. Setting this parameter too high may slow your system. See also DRAM, RAM, Refresh, and SRAM.

WAN, Wide Area Network: A network connection between two locations, typically a T-1 circuit, but may be ISDN, Frame Relay or other (possibly a VPN over any Internet connection type) used to bridge Local Area Networks in related office facilities.

War-chalking, war-walking, war-driving: The activities of surveying an area looking for wireless network hotspots and access points, then marking the direction and type of services available in chalk on sidewalks or walls. Derived from the war-dialing of "War Games" movie fame.

WEP, Wired Equivalence Privacy: A scheme used to make the data traveling on wireless networks unreadable by those not authorized to use your network. 'Wired Equivalence' indicates the scheme proposes to make wireless signals as relatively secure from intrusion as using a wired system. Both the 64- and 128-bit WEP encryption schemes can be deciphered by commonly available software tools so WEP is not to be trusted for secure, valuable or private data. Consider a VPN solution to help secure your wireless network.

Windows: A Microsoft multitasking and graphical user interface that allows multiple programs to operate on the same PC system and share the same resources.

Windows NT: A Microsoft 32-bit multitasking operating system and graphical user interface.

WLAN, Wireless LAN: Inter-connections between client computers, servers and other devices over radio waves versus Ethernet cabling connections.

Workstation: A user's computer system attached to a network. Workstations do not necessarily contain diskette or hard disk drives, instead using built-in programs to boot-up and attach to a network server, from which all programs and data files are obtained.

WPA, Wi-Fi Protected Access: A pre-802.11i wireless LAN security method.

Write Protected: The status of a diskette with a write-protection tab or slot. All 5-1/4 inch diskettes use a write-protect notch and stick-on tab to define write protected status. If the notch is covered, the disk is write-protected. All 3.5 inch diskettes use a sliding window cover over a small hole in the near left corner of the casing (shutter door facing away from you). If the hole is uncovered, the disk is write protected.

WWW or World Wide Web: A term used to describe multiple inter-networked computer systems providing text and graphical content through the HyperText Transfer Protocol (HTTP), usually over Internet Protocol (IP) networks.

XMS (Extended Memory Specification): A standard that defines access and control over upper, high, and extended memory on 286 and higher computer systems. XMS support is provided by loading the HIMEM.SYS device driver or other memory management software that provides XMS features.

XT: The second model of IBM PC series provided with 'extended technology' allowing the addition of hard disks and eight add-in card slots. The original XT models had between 64 k and 256 k of RAM on board, a single floppy drive, and a 10 MB hard disk.

Index

A

AAG weather station, 89f, 91f
AAG weather station sensor
  antenna mast, 93f
access-point without router/firewall, 33
Afterbumer products, 33
AGW Packet Engine (AGWPE), 71
  icons, 74f
  before port configuration, 75f
  software, 72
  successful port configuration, 75f
  TCP/IP drivers, 72, 73f
AGWTCPIP.INF file, 74f
AirLink, 32
Amateur Position Reporting System (APRS), 69–82
  GPS data mapping program, 81
  Internet connectivity features, 83
  parts list, 71
  Plus, 72
  +SA, 77f–78f
  steps, 72–80
  tools, 72
amateur radio (ham) operators, 69
APC Biopod, 137
APRS. see Amateur Position Reporting System (APRS)
APRS Plus, 72
APRS+SA
  first location, 78f
  Internet settings, 77f
  settings, 77f
  setup, 77f
audio connections, 6
  cable between PC sound card and tuner card, 97f

B

bad power, 23
Barbie PC, 1
BBC Climate Change Experiment, 124
Belkin, 32
Berkeley Open Infrastructure for Network Computing
      (BIONC), 122, 124f
BOINC Manager program, 124f
built-in router/firewall, 33
bundle security software, 136
Bux Comm Corporation, 78
Bux Comm RASCAL

handmade interface cable, 78f
PC-to-Radio interface module, 78f

C

cables
  dressed and bundled, 14f
  dressed-up, 13
caller-ID, 113
car-PC, 83
CD-ROM drive
  faceplate, 3f
cesium-reference clock, 128f
channel selections
  labels, 100f
chassis temperature sensor, 7, 8f
Cingular's General Packet Radio Service, 64
  active Internet connection, 66f
  wireless card, 66f
Cisco, 109
Citizen Weather Observer Program (CWOP), 69, 90
  Weather Engine upload service, 94f
cleaner electrical power, 23
climateprediction.net, 124
cold-cathode lamps, 10
computer security, 135–141
  devices, 139–140
configuration page, 32f
connector bezel
  mounted on rear panel, 20f
control panel, 5–8, 6f
  cables, 7f
  installing, 7f
  parts list, 5
  steps, 6
  tool kit, 5
Corcom 10VR1 (Mouser.com), 24
CPU temperature, 8f
Create Fingerprint logon, 139f
creativemods.com, 15
CWOP. see Citizen Weather Observer Program (CWOP)
cyberguys.com, 1f, 111
cyberlink.com, 52

D

data-over radio packet modem, 70f
Davis Instruments, 89

DC adapter, 53f
DC-to-AC adapter, 53f
DeLorme's Street Atlas, 59, 60
Device Lock form, 140
Digital Persona
    fingerprint-enabled web pages, 140f
    web-site logon, 140f
Digital Persona U, 137f, 138f, 139f
DIMES, 122
disk drives
    connected to new motherboard, 22f
    install, 13
distributedcomputing.info, 125
Distributed Internet Measurements &
        Simulations (DIMES), 122
D-Link, 32
    access point, 34f, 35f
dynamic picture frame, 39–49
    parts list, 40
    steps, 41–48

E

Earth Bridge, 88
    GooglEarth, 85f
    GPS and satellite reception status, 85f
    preference tab, 85f
Earth.NetDIMES, 121
earthquake straps and brackets, 57f
EchoLink, 109
Einstein Home, 121, 122, 123f
11b WiFi products, 32f
11 Mbps, non-WPA 802, 32f
EV-DO wireless PC data card, 66f
Evil Genius VoIP phone patch, 112
Evolution Data Optimized (1X EV-DO), 64
existing hard drive
    replacing, 17
expandable sleeves, 13
external grille, 11

F

ffan, 6, 11
    mountings, 12f
fantasy-land, 9
fiberglass
    creating, 15
filler panel, 6f
filter, 25
FindU.com, 93f
fingerprint readers, 136
fingerprint registration, 137f
54 Mbps, WPA-compatible 02.11 g WiFi products, 33f
FM modulator, 53
Folding Home DNA, 121, 123f
Folding.stanford edu, 125
Franson, Johan, 71

Franson GPSGate software, 72, 84
    input port configuration, 81f
Franson Technology AB, 71
front panel switches
    connections, 21f

G

Garmin, 59
General Packet Radio Service (GPRS), 64
    active Internet connection, 66f
    wireless card, 66f
Generation Y, vii
generic PC-to-radio interface cable
    schematic, 79f
GeoCities web page, 84
GEtrax, 83
    software, 84
Global Positioning System (GPS), 59
    hand-held, 59, 59f
glossary, 143–168
glow-in-UV-light cables, 13
GooglEarth
    Earth Bridge, 85f
Google Earth KML file, 88
Google Earth program, 83, 84
    GPS/car-PC, 87f
    KMI file, 86f
    update track, 87f
Google file, 83
Google's Picasa
    building slide show, 44f
Google Tracker, 88
Google tracking trip, 83–88
    parts list, 84
    steps84
GPRS. *see* General Packet Radio Service (GPRS)
GPS. *see* Global Positioning System (GPS)
GPSDIAG program, 62f, 63f
GPSGate, 71
    output port options, 82f
green frog mouse, 1f
green PC, 1–4
    completed, 4f
    parts list, 2
    steps, 2
    tool kit, 2
grille, 12f
grommets, 28f

H

ham operators, 69
hand-held GPS, 59f
hand-held GPS5, 59
hand-held-radio
    Kenwood TH-D7, 72
handmade Rascal-to-Icom radio cable, 79f

headphone splitter, 53
HighSpeed Downlink Packet Access (HSDPA), 64
HSDPA. *see* HighSpeed Downlink Packet Access (HSDPA)
Hyco bushings, 26, 28f

I

IIcom portable radio, 78f
image/build slide show
    selecting, 41f
in-car routing devices, 59
Internet mapping sites, 70
Internet Radio Linking Project (IRLP)109, 109
Internet Time dialog, 129f
intervideo.com, 52
InterVideo's WinDVD (www.intervideo.com), 52
IRLP. *see* Internet Radio Linking Project (IRLP)109

K

Kantronics KPC-3, 70f
Kenwood TH-D7 hand held-radio, 72
keyboard
    supplier, 1
keystone button
    cutting off, 20f
kid-friendly theme, 4
KipSSPE, 72
    middleware program configuration, 76f

L

LaCrosse Technology, 89
laptop
    audio output, 54f
    DVD player testing, 54f
    FM modulator connection, 54f
    mount, 5f
LCD display
    painting, 3f
    removing masking tape, 4f
LED indicators
    connections, 21f
LHC@home, 125
Linksys, 32
local administrator password, 34f
location
    marking current, 63f
Logitech camera, 115f

M

Magellan GPS receivers, 59
Matrix-black-techno-look PC makeovers, 40
Mbps, non-WPA 802, 32f

Mbps, WPA-compatible 02.11 g WiFi products, 33f
metal-oxide varistor (MOV), 23
Microsoft Fingerprint Reader, 137
Microsoft Knowledge Base, 140
Microsoft Photo Story
    selecting slides, 44f
Microsoft Streets and Trips, 59, 62f, 83
Microsoft Streets and Trips with GPS locator, 60
MIMO (multiple-input/multiple-output), 31
mobile entertainment, 51–56
    parts list, 51
    steps, 52–56
mobile navigation, 59–64
    parts list, 66
    steps, 66
motherboard
    aligning new, 19f
    install into chassis, 13
    nylon stud supporting, 20f
    removing from mounting plate, 19f
    replacing, 17
    upgrade, 17
        parts list, 17
        steps, 18
        tool kit, 17
mount
    final pieces, 55f
    PC playing DVD, 57f
mounting base, 56f
    positioning, 54f
mounting plate
    removing from C chassis, 19f
mounting post and table installed, 56f
mouse
    supplier, 1
Mouser.com, 24
MOV, 23
multi-function PC control panel, 5
multiple-input/multiple-output, 31
musical picture frame, 43

N

National Bureau of Standards (NBS), 127
National Institute of Standards and Technology (NIST), 127
NBS. *see* National Bureau of Standards (NBS)
NetDIMES
    data paths, 124f
    Internet mapping project, 122
netdimes.org, 125
Netgear, 32
network authentication, 37
Network Time Protocol (NTP), 127
    servers, 129
NIST. *see* National Institute of Standards and Technology (NIST)
Northern Telecom, 109
Novell NetDrive, 84
NTP. *see* Network Time Protocol (NTP)

O

Oregon Scientific, 89

P

packet modem
  cost, 70
password, 135
  local administrator, 34f
PC. *see* personal computer (PC)
pentium D dual-core processor, 17
personal computer (PC)
  chassis, 10
  control panel, 5–8
    parts list, 5
    steps, 6
    tool kit, 5
  front cover, 3f
  science projects, vii
  tuner and video capture I/O card, 97
  TV, 97–100
    parts list, 98
    steps, 98–99
    viewing screen, 100f
  ventilation, 7
  weather station, 89–95
    parts list, 90
    steps, 91–94
    tool list, 90
    warnings, 90
personal firewalls, 135
picture frame laptop
  completed, 48f
Pinnacle Systems PCTV card, 97
plastic creatures
  creating, 15
playlist
  saving, 46f
power and data cables
  disconnected from old motherboard, 19f
PowerDVD (www.cyberlink.com), 52
power filter
  holes, 26f
power filter and fuse holder
  mounting, 26f
power line noise filter and protection device, 23
power protection
  build your own, 23–29
    parts list, 24
    schematic, 24
    steps, 25–28
    tool list, 24
power surges, 23
Predictor@home, 125
preferred networks list, 37f
pre-molded cases
  suppliers, 15

Q

Quantum Monte Carlo at Home, 125

R

radio-to-interface cable, 72
radio-to-PC interface unit, 72
RAM
  increasing, 17
  installing additional, 52f
RASCAL radio-to-sound card interface unit, 78
Rascal-to-Icom radio cable, 79f
real-time clock circuit, 127
redecorated PC, 4
Rosetta@home, 125
Rossopoulos, George, 71
RoutePlanner, 63
route planner, 64f

S

SATA. *see* serial-ATA (SATA) interface
science explorations, 121–124
Search for Extra-Terrestrial Intelligence (SETI), 122
  Home project, 121
  project, 122f
Seasonal Attribution Project, 125
Secure Wave Sanctuary Device Control, 140
security webcam, 115–119
  parts list, 115
  steps, 116–118
serial-ATA (SATA) interface, 18
serial-to-USB adapter, 61f
SETI. *see* Search for Extra-Terrestrial Intelligence (SETI)
SETI@home, 125
SETI.ORG, 122
side panels, 11, 12f
SIMAP, 124
Skype, 109
  completed setup, 110f
  e-mail address, 110f
  setup screen, 110f
Skype-to-Skype, 111
Skype VoIP
  phone call, 110f
SLEEP.EXE, 84
slide show
  creating, 43f
Slingbox, 101f
  accepting input, 107
  adjusting remote control settings, 104f
  assigning passwords, 104f
  configuration information, 106f
  configuring to stream Internet connection, 104f
  connections, 102f
  Finder ID, 106f

Index

firewall/router configuration, 105f
manually assigning TCP/IP parameters, 105f
naming, 104f
network appliance, 101
picture quality and sound level, 102f
properties, 106
selecting automatic or manual network settings, 105f
selecting manufacturer, 103f
signal feeds, 103f
signal inputs, 102f
SlingMedia, 101
SlingPlayer
    first installation screen, 102f
    IP port, 105f
    program, 101
    software, 101
sling video, 101–107
    parts list, 101
    steps, 101
Smart Beaconing, 83
Smart Beacon parameter list
    APRS+SA, 80f
Smart Beacon settings
    APRS+SA, 80f
Softex OmniPass software, 137
Sony SNC-RZ30N Network Camera, 119f
sound inputs and volume levels, 99f
sound recording input, 99
sound recording level, 99
spyware, 135
SSID
    renaming access point, 34f
Sunbeamtech control panel, 6f
Sunbeamtech 20-in-1 superior panel, 5
surge-protection outlet strip, 23
SV2AGW CPIP Over Radio NDIS Driver, 74f
SV2AGW drivers, 74f
S-video sources, 97f
SwiftWx (www.swiftwx.com), 67
sync, 127–134
    brief history of time, 133
    PC time accuracy with time server, 130–132
SZTAKI Desktop Grid, 125

T

TAI8515 test, 90
Tanpaku, 124
Tardis
    adjustments, 132f
    COM port speed, 133f
    configuring for local time server, 133f
    diagnostic information, 132f
    GPS receiver, 132, 133f
    installation, 131f
    NTP servers, 131f
    server, 131f
TCP/IC driver, 73f
TCP/IP over radio

    enabling, 76f
techno-clutter, 5
techno-PC
    new style, 9
temperature sensor, 8f
thermaltake.com, 15
ThumbsPlus, 41
    setting filename, 42f
    slide show, 40f
time-keeping software, 127
Tiny Tracker 3 packet modem, 70f
TKIP
    data encryption, 37
tuner and video capture I/O card, 97
tuner cards, 97f
TV, 97–100
    channel selections, 99f
    parts list, 98
    playback input control99
    steps, 98–99
    tuner cards, 97
        installing in open PCI slot, 97f
    tuner configurations program, 99f
    viewing screen, 100f
two-headphone splitter, 53f

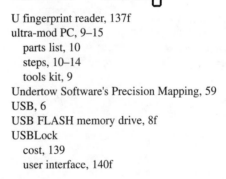

U

U fingerprint reader, 137f
ultra-mod PC, 9–15
    parts list, 10
    steps, 10–14
    tools kit, 9
Undertow Software's Precision Mapping, 59
USB, 6
USB FLASH memory drive, 8f
USBLock
    cost, 139
    user interface, 140f

V

Velcro pads
    holding laptop on mobile table, 56f
ventilation, 7
Verizon's IX EV-DO wireless PC data card, 66f
video capture I/O card, 97
video recording file compression type and location
    saving recorded shows, 99f
video source
    selecting for tuner card, 99f
virus protection, 135
voice-over IP (VoIP)
    not emergency-ready, 111
    phones, cost, 111
    with Skype, 109–113
        parts list, 109, 111
        phone patch, 111

steps, 110, 112
Vonage, 109

W

Weather Engine 5, 90
Weather Observer Program, 68
weather station, 89–95
   parts list, 90
   serial/COM port adapter plug, 91f
   software configuration, 92f
   steps, 91–94
   tool list, 90
   warnings, 90
weather tracking, 67
Weather Underground, 93f
Weather Underground upload service
   Weather Engine, 93f
webcamXP
   configuring FTP upload, 117f
   cost, 115
   monitoring and time-stamping, 117f
   option list, 116f
   selecting recording options, 118f
WEP. *see* Wired Equivalency Protocol (WEP)
WiFi products, 32f
WiFi secure, 31–37
   parts list, 33
   steps installing new access point, 34
   steps installing new WiFi card, 35–36
Windows Device Manager, 60
   COM ports, 61f, 75f

Windows Media Player
   creating playlist, 46f
   playlist files, 45f
   repeating playlist, 47f
   starting new playlist, 45f
Windows Media Video (WMV), 41
Windows Network Connections, 36f
   access, 36f
   available adapters, 36f
   security settings, 37f
Windows Remote Desktop, 47
   controlling picture frame laptop, 48f
Wired Equivalency Protocol (WEP), 31
wired or wireless connected piece of art, 39f
Wireless Protected Access (WPA) security, 31
wires, 7f
1-Wire Weather Engine, 93
1-Wire Weather Instrument Kit, 89
WMV. *see* Windows Media Video (WMV)
World Community Grid, 124
WPA. *see* Wireless Protected Access (WPA) security
www.creativemods.com, 15
www.cyberlink.com, 52
www.intervideo.com, 52
www.netdimes.org, 125
www.swiftwx.com, 67
www.thermaltake.com, 15
www.xoxide.com, 15

X

xoxide.com, 15